Tire Engineering

Tire Engineering

An Introduction

Brendan Rodgers

CRC Press
Taylor & Francis Group
Boca Raton London New York

CRC Press is an imprint of the
Taylor & Francis Group, an **informa** business

First edition published 2021
by CRC Press
6000 Broken Sound Parkway NW, Suite 300, Boca Raton, FL 33487-2742

and by CRC Press
2 Park Square, Milton Park, Abingdon, Oxon, OX14 4RN

© 2021 Taylor & Francis Group, LLC

CRC Press is an imprint of Taylor & Francis Group, LLC

ISBN: 978-0-367-44228-6 (hbk)
ISBN: 978-1-003-02296-1 (ebk)

Typeset in Times
by Deanta Global Publishing Services, Chennai, India

Visit the [companion website/eResources]: [insert comp website/eResources URL]

Contents

Preface

Tire technology is a complex combination of materials science, mathematics, and structural engineering. There is a wide body of literature covering the topic of rubber chemistry and technology but, given the size of the tire industry and its global footprint, there is surprisingly very little available on the subject of tire engineering. This text was, therefore, largely inspired by Fred Kovac's book on Tire Technology published in 1978. Since Fred Kovac, a former Vice President for Technology at The Goodyear Tire & Rubber Co, published his text, very little has emerged in the public domain. One reason for this is the broad diversity found in the industry. For example, one set of technologies that is implemented by one manufacturer may not be readily adaptable by another. What works for one manufacturer seldom works for another. Apart from patent, trade secret, and proprietary constraints in adapting different technologies, other major hurdles include differences in compound line-ups, raw material quality differences, sources of supply, tire design and construction, and manufacturing operations; everyone is different. Even within the same company, it can be challenging to manufacture the same tire construction at different plants because, when the plants were built, differences arose with respect to compound-mixing operations, component preparation equipment, tire building or curing. Though the individual influences might be small, local plant operating procedures may also impact the final quality of a tire enough to affect Original Equipment qualification or other performance parameters.

Given the variables in tire design, construction, and manufacturing, there is still a broad range of design principles common to all tires. The modern radial tire can still be considered in two sections, the tread or crown region and the casing. Between tire lines, such as passenger car tires and light truck tires, there are many common design features and compounds. Furthermore, the materials can be processed on the same equipment, simplifying some manufacturing operations.

This text, therefore, focuses on common principles. It is intended only as an introductory handbook for the industrial and development engineer and not as a theoretical treatise. It is based on much practical experience in a broad range of global tire operations from materials to design, manufacturing, and applications. In addition, it is intended to be holistic, in that design, construction, materials, and performance engineering are all covered. The focus is on practice rather than theory. Furthermore, each chapter is designed as a stand-alone review, with a glossary at the beginning of the text. Given the differences in design and manufacturing philosophies among all of the global tire manufacturers, it can offer only one perspective, most likely quite different from that of design engineers exposed to only one manufacturer. However, as a reference or troubleshooting guide, it is hoped the reader will find the text of considerable value.

Looking forward, of the many future trends discussed in the text, two, in particular, merit comment. First, governments will continue to support the tire industry because of the creation and need for direct and indirect labor and, more importantly, skilled labor. As the industry automates its manufacturing operations, the demand for skilled labor will only increase. Second, is the broad area of sustainability. This has the potential to move from a broad set of general guidelines and talking points to a hard engineering discipline, focusing on tire factory water and power consumption, reductions in emissions, and the near-elimination of post-industrial waste. Along with increasing automation, successful implementation of a well-defined sustainability program could therefore lead to improved production efficiencies, quality, cost reductions, and profitability. The industry, therefore, has a potentially strong future.

The compilation of a work such as this, albeit at an introductory level, is not possible without the help of many people. I am especially indebted to Dr James P. Stokes, Polymers Technology Manager at the ExxonMobil Chemical Company for his support over the past 18 years at ExxonMobil, to Kho Irani and Justin Grafton, also at ExxonMobil's Polymer Technology Department, and to Derek Kato at ExxonMobil's Law Department for his comments, to Phillip Boltan at the University of Texas in Austin for his help with the manuscript and to Allison Shatkin and Gabrielle Vernachio at CRC Press for their patience and support.

Author Biography

Brendan Rodgers was a Technology Advisor at the ExxonMobil Shanghai Technical Center, having moved to Shanghai at the end of 2014 from Texas, returning to Houston in 2018. He worked with ExxonMobil's Butyl Technology and Specialty Polymers teams, supporting programs for the tire industry. Before moving to Shanghai, he was a development engineer at the ExxonMobil Baytown Technology Center, working on applied nanocomposite technology, web-based information systems, and introduction of new technologies in tire manufacturing in Asia and Saudi Arabia.

Before joining ExxonMobil in 2002 he was an engineer at The Goodyear Tire & Rubber Company in Akron, Ohio, and led teams working on truck tire, farm, off-road, and aircraft tires. He also had assignments in Ireland, Italy, and Luxembourg, working on original equipment automobile tires, truck tires, industrial rubber products, and new tire materials technologies. He is the originator of a broad range of patents in tire technology, and authored many industry publications, including being Editor of the text, *Rubber Compounding, Chemistry and Applications.*

Brendan has a PhD in chemical engineering from The Queen's University of Belfast in Northern Ireland, studying thermodynamics, heat transfer through large rubber sections, and vulcanization kinetics. He has a Master's degree in Polymer Technology, and a BSc in Chemistry.

Glossary

Many terms used in the tire industry have been explained and defined in the text. There are also a broad range of additional terms requiring a concise definition, and these have been consolidated below.

Ackerman angle	The relative angle between the left and right steering axle wheels during a turn. This compensates for the difference in turning radius between the two front wheels, thereby minimizing lateral tire scrubbing and wear.
Ackerman principle	Alignment principle, based on vehicle steering axle wheels and alignment upon which the turning angle is calculated
Adjustments	Tires prematurely removed from service.
Alignment	Steering and suspension adjustment to facilitate operation of tire and wheel assemblies for optimum vehicle control and tire wear.
Apex	The triangular component above the bead and extending approximately 25–30% up the sidewall length. It is also referred to as the bead filler and, depending on the tire size, can be just one compound (passenger car tire), or two compounds, termed Apex 1 and Apex 2, as in truck tires. The design function is to provide a modulus gradient up the sidewall, thereby
Aspect ratio	Section height divided by section width, expressed as a whole number. A conventional-sized tire has a nominal aspect ratio of "80", i.e., the section height is 80% of the section width.
Asymmetrical	Tread design where the tread pattern on one side of the tire centerline is different from that on the other side.
Axle	Beam running crosswise, supporting the chassis.
Axle, live	Drive axle.
Axle, pusher	Non-powered rear axle.
Axle, rating	Maximum load which the front and rear axles are designed to support.
Axle, tag	Non-powered rear axle behind drive axles (free rolling but helps distribute total vehicle weight).
Axle, tandem	Two axles paired together, which may or may not be powered.
Axle weight	Weight carried by the tire on one axle.
Axle yaw	Oscillation of an axle about the horizontal transverse through the center of gravity, which could induce irregular tire wear.
Balance	The distribution of weight around the circumference of the tire and wheel assembly.

Bead	The region of the tire designed to lock the tire onto the rim. The bead cable is made of spiraled wire, covered by hard rubber compound. It also holds the ply in place. The bead or lower region of the tire consists of the following components: 1) steel wire bead ring, which locks the tire onto the wheel or rim; 2) apex or bead filler; 3) a fabric insulation layer, sometimes called a "flipper", to prevent abrasion of the chipper and bead wire; 4) chipper wire, which protects the ply from rim damage; and 5) the chafer, which is the abrasion-resistant rubber component in contact with the rim. The apex or bead filler compound are formulated for dynamic stiffness, enable stress distribution in the lower sidewall region, and provide good tire–vehicle handling properties. The bead insulation compound must have good steel bead–wire adhesion and also bead–ply compound adhesion. The chafer or rim strip compound protects the plies from rim abrasion and helps seal the tire to the rim.
Barrier	Also termed a squeegee or tie gum and provides an adhesive layer and reinforcement layer for the inner liner.
Bead seat	Flat portion of the rim on which the tire bead sits.
Bead wire	Coil of high tensile strength wire, either brass-coated or bronze-coated. Each wire is coated with a thin layer of rubber for adhesion.
Bead wrap	Fabric wrap to hold the bead cables together during tire manufacture.
Belts	Layers of steel wire, and sometimes fabric, that form a hoop under the tread. They restrict deformation of the casing, provide rigidity to the tread region, and provide a stable foundation for the tread components. This allows improved wear performance, vehicle stability, handling performance, damage resistance, and protection of the ply cords. Produced using brass-coated wire.
Belt wedges	Components placed between adjacent belts, at the belt edges, to help maintain a flat footprint when the tire is under load. In terms of nomenclature, an example is the 2/3 wedge in a truck tire, which is the wedge at the end of Belt #2 and Belt #3, stabilizing the belt ending.
Bias tire	A tire where the ply cords are at an acute angle to the circumferential centerline of the tire. The angle would typically be of the order of 36° but will vary depending on the tire manufacturer.
Butyl rubbers	Polymer used in tire inner liners to facilitate inflation pressure retention.
Camber	Tilt of the front wheels. Inward at the top is negative, outward at the top is positive.
Casing	Term used interchangeably with "carcass", and is the body of the tire, excluding the tread region.
Caster	Amount of tilt in the front and rear of axis of the axle beam.

Chafer	A hard rubber compound which helps lock the tire onto the rim. It can also be referred to as a toe guard (passenger car tires).
Chip/chunk/cut	Cutting, tearing, and breaking-off of tread pieces, due to rough road conditions.
COE	Cab Over Engine chassis. Contrary to this is "conventional", which the term used to describe when the cab is set back from the engine, rather than on top of it.
Coefficient of friction	Horizontal force required to move an object divided by its weight.
Compound	Blend of elastomers and chemicals, vulcanized to meet a set of defined mechanical properties.
Compounding	One of the most complex fields in materials science, and including polymer science and technology, formulating and blending, to produce compositions that meet the mechanical properties needed for a tire, thereby enabling it to perform to its design objectives.
Conicity	Tire imperfection causing the tire to adopt the shape of a cone when inflated and loaded, resulting in the creation of a lateral force.
Contact area	Total pavement contact area of a mounted and loaded tire tread. "Gross" is the total contact area and "Net" excludes the grooves and other voids.
Contact pressure	Tire footprint pressure distribution.
Cord	Strands of wire or textile forming the ply or belts.
Cord separation	Cord parting away from adjacent rubber compounds.
Cornering coefficient	Lateral force exerted on a tire divided by the vertical load at a defined slip angle.
Crown	Tread region of the tire construction. Crown radius is the measure of the curvature, or an indication of the tread flatness. Crown width is the distance, shoulder to shoulder
Cubic capacity	Carrying capacity of a vehicle, measured by volume. As truck and trailer volume increases, smaller tires are needed, due to lower platform heights. The platform must be lowered as the roof cannot be raised due to bridge heights.
Cushion	The rubber layer between the tread base and the tire belt assembly.
Damping	The ability of a rubber compound or tire construction to absorb energy and reduce vibration. Related to hysteresis.
Deflection	The difference between the loaded and unloaded tire section heights at a defined load and inflation pressure. Related to "Spring rate".

Denier	The weight of cord, expressed in g per 9000 meters.
Differential	Rear axle gear assembly, enabling one axle shaft to turn slower or faster than the other, while turning or maneuvering.
Drivetrain	Clutch, transmission, driveline, and drive axles.
Duals	Pair of tires mounted together, typically on drive and trailer axles.
Dual axle	Two axles in combination.
Dual spacing	Distance between the centerline of one dual-mounted tire to the centerline of the second tire. It is important for dual-mounted tires on heavy-duty trucks to prevent adjacent tire sidewalls from abrading against each other. It is determined by adding two offsets on disc wheels or two offsets plus one spacer disc width for demountable rims.
Durability	Laboratory road wheel (1.7 m diameter) test time to removal.
Dynamometer	Hydraulic- or electric-powered rollers designed to measure the output of drive train friction.
Endurance	A determination of a tire's service life to removal.
EPI	Ends of cord per inch (25.4 mm) width of fabric.
Equilibrium temperature	Maximum temperature generated when a tire is in operation, when heat generation is equal to heat dissipation.
Fifth wheel	Coupling device used to connect a trailer to a tractor. It is a steel coupling device, bolted to the truck chassis, to engage the trailer kingpin. This permits the trailer to pivot when the tractor–trailer combination turns, maneuvers, or changes direction.
Filling	Light threads that run at right angles to the warp (also referred to as the "pick") that serves to hold the fabric together.
Flange height	Rim flange height above the rim base.
Flipper	Fabric, located between the bead and the ply, preventing the two components from damage due to abrasion as the tire rotates under load.
Footprint	The section of the tread in contact with the pavement. An optimized footprint is essential for good tread wear.
Footprint length	Length of the loaded footprint.
Footprint width	Width of the loaded footprint.
Force and moment	Describes the forces acting on a tire during cornering.
Force variation	Periodic variation in vertical forces of a free-rolling tire, repeated with each revolution.
Gough stiffness	Lateral stiffness of the belt system (the composite including the reinforcement and compounds). Bending moment in this instance is the force and distance from the point of support to the point of force application, causing bending and distortion.

GAWR	Gross Axle Weight Rating: Lowest-rated member from the components of tires, suspension, wheels and rims, beams, and brakes.
GCW	Gross combination weight; total weight of truck, tractor, and trailers plus payload.
GCWR	Gross Combination Weight Rating: Manufacturer's rating for maximum allowable weight of truck or combination of truck and trailers.
GVW	Gross vehicle weight; total weight of fully equipped truck and payload
Governor	Device to control engine maximum rpm.
Green strength	Tensile strength of a rubber compound before it has been vulcanized.
Green tire	Tire before it is cured or vulcanized in a mold.
Gross contact area	Total footprint area.
Gum strips	These are different types of strips, all of the same compound, located at belt endings, ply endings, and other component endings, which serve as a transition compound between two different tire parts, providing a modulus gradient, improving component-to-component adhesion, and minimizing potential for crack formation.
GVW	Gross vehicle weight; total weight of fully equipped truck and payload.
Harmonic	Periodic or rhythmic force variations in a sinusoidal manner around a tire. One phase is described as a 1st harmonic. When two phases are noted, it is described as a 2nd harmonic. Lateral force variation 1st harmonic is typically due to a tread splice. Radial harmonic may be due to irregular placement of the belt lay-up.
Heel-and-toe wear	Uneven sloping wear across tread elements. It can cause a shifting movement of tread elements as the tire rotates through the footprint.
Hydroplaning	Loss of traction on wet surfaces, at high speeds.
Impact break	Diagonal split or breaks in the shape of an X, visible from the inside of the tire.
Inner liner	Membrane consisting of compounded low-permeability rubber such as bromobutyl rubber (BIIR) or chlorobutyl rubber (CIIR) and whose function is to retain compressed air inside the tire. The liner typically reaches from the bead on one side of the tire to the opposite side, thereby providing a seal. High-quality halobutyl inner liners are critical to the operation of redial tubeless tires. It is, therefore, a critical strip of rubber on the inside of the tire, ensuring that the tire maintains its inflation pressure.
Kingpin	Front steering axle pin, allowing wheels to turn and to steer the vehicle.

Lateral force Variation or change in force from one side of the tire to the other as it rotates.

Lateral force coefficient The lateral force divided by the vertical load.

Lateral runout Difference between the maximum and minimum measurements parallel to the spin axis at the widest point of each tire sidewall when the tire is mounted on a wheel.

Load distribution Distribution of load on a vehicle.

Load range Letter designating the load-carrying capability of a tire.

Lock ring Third piece of a 3-piece rim assembly, locking the side ring to the rim base.

Loaded section width Width of a loaded tire section at its rated inflation pressure.

Longitudinal force Force acting in the direction of travel while accelerating, or opposite to the direction of travel during braking, or rolling resistance.

Low-profile radial Tire aspect ratio less than 80, i.e., 75 and lower. Aspect ratio of 80 and above are considered standard profile (sometimes referred to as standard section).

Lug Tread blocks running across a tread pattern rather than a solid circumferential rib.

Multi-piece rims Rim consisting of two or more parts. Two-piece rims have a base and side ring. Three-piece rims consist of a rim base, side ring, and lock ring.

Net contact area Area of tread rubber compound in contact with the pavement. Excludes voids.

Natural rubber Primary compounding material in radial tires, due to its high tensile strength, tear strength, and low hysteresis.

Net-to-gross contact area Footprint area excluding the tread voids, grooves, or space between lugs.

On-/off-road Tires for operating on vehicles travelling on both paved roads and off-roads, such as lumber haulage, coal mining, farming, and trash haulage. Such tires have special tread designs and tread compounds.

Overall diameter Mounted and inflated but not loaded new tire diameter.

Overlay Calendared fabric layer, alternatively applied as strips, above the belts and under the tread. This fabric layer is typically made from nylon 6,6 (preferred) or nylon 6, lies between the belts and the tread, and is intended to constrain centrifugal forces when the tire is rotating at high speed. It may be full width, spiral wounded fabric-reinforced strips, or belt edge cover strips.

Oversteer Tendency for the rear axle of a vehicle to lose grip before the front axle.

Payload	Rated load-carrying capability of a vehicle, typically defined by the original equipment manufacturer.
Peak traction coefficient	Ratio of peak traction to the vertical load.
Peak traction	Maximum breaking force before the wheel locks.
Percent hydroplaning	Dynamic footprint area divided by the static footprint area and expressed as a percent.
Pick cords	Light threads that run at right angles to the warp (Filling). This helps to maintain fabric integrity
Pitch length	Length of the individual tread pattern element circumferentially around the tread.
Plies	Ply cords consist of steel or fabric, extend from the bead on one side of the tire to the opposite side, and serve as the primary reinforcing member on the tire. They provide the strength to the tire, that enables it to retain air at high pressures. Automobile radial tires can have one or two layers of polyester cords, whereas bias truck tires can have many more.
Ply line	Geometrical cross-section curve at rated inflation pressure.
Powertrain	Engine plus drivetrain.
Radial	Tire where the ply cords run bead to bead, 90° to the centerline of the tire or to the direction of rotation.
Radial force	Force acting on a tire perpendicular to the centerline of rotation or direction of the axle. It is caused by heavy tire component splices and will increase radial force. Soft spots in the tire, such as that due to stretched ply cords, cause a decrease in radial force. Radial force variation is a summation of the radial 1st, 2nd, 3rd, etc. harmonic.
Radial runout	Difference between the maximum and minimum measurements on the tread surface and in a plane perpendicular to the spin axis, while the tire is mounted on a wheel. It is a measure of the out-of-roundness of the tire. It can also be termed 'centerline runout'.
Regrooving	Cutting the tread pattern deeper than the original design. Typically extends down to the start of the under-tread.
Resilience	Ratio of energy output to energy input (hysteresis). Whole-tire hysteresis reflects whole-tire rolling resistance, i.e., energy loss.
Retreadability	Number of times a tire can be safely retreaded.
Revolutions per mile	Number of revolutions a tire will make in one mile, i.e., rpm (alternatively, revolutions per kilometer).
Rib	Tread section running circumferentially around the tire.

Rim   Metal support wheel for the tire, produced from either steel or aluminum.

Rim width   Distance between rim flanges (frequently defined in inches), and is defined for each size of tire.

Rim flange   Vertical collar on the rim, which retains the outer edge of the tire bead.

Rivet   Distance between cords in a fabric; high rivet typically describes a construction with wider distances between cords. A low rivet indicates high ply or high belt cord/wire cord density.

Rolling radius   Distance from the tire center to the pavement at rated capacity.

Rolling resistance   Resistance of a tire to rolling. It has a direct impact on vehicle fuel economy and is influenced most by compound hysteretic properties.

Scorch   Pre-vulcanization of compound caused by processing at high temperatures. It can occur as small cured particles or lumps of compound in the extrusion or calendaring operations.

Sealant   Component designed to block an air leakage in the event of a puncture.

Section width   New tire section width, excluding scuff ribs and lettering.

Section height   Distance from the tire centerline to the rim.

Self-aligning torque   The stabilizing reaction to slip angle which helps the tire and vehicle to return to neutral conditions at the completion of a maneuver.

Shoulder   Outer edge of the tread and upper sidewall.

Shoulder wedge   A stiff elastomeric component located at the edge of the belt assembly and designed to provide stability to the shoulder region of the tire, provide a uniform footprint, and help dissipate heat. Triangular-shaped component located at the edge of Belt #1 and between the ply and sidewall.

Sidewall   Provides long-term weathering protection and casing durability. It also protects the tire from impact and curb-scuffing. It is compounded to resist fatigue and flex-related cracking.

Sidewall insert   Compounded rubber or a calendared fabric strip, extending from the bead region up the sidewall.

Sipes   Indentations in the tire tread designed to increase tire traction performance.

Slide traction coefficient The ratio of slide traction forces to the vertical load.

Slide traction   The force developed during breaking after the wheel has locked.

Slip angle   Angle between the vehicle's direction of travel and the direction in which the front wheels are pointing.

Slots	Narrow grooves in new tire tread patterns. They are produced by inserts in the tire mold (tread segments) They are also referred to as sipes.
SMR	Standard Malaysian Rubber.
SRP	Serum rubber particles.
Speed rating	Alphabetic ratings which define the design speed capability of the tire. The letter is incorporated into the size description of the tire. For example, a 195/75SR14 has a speed rating of 'S'. Tables of speed ratings and corresponding alphabetic designations are published by the Tire & Rim Association.
Spinout	When maximum drive slip exceeds resistance as a result of tractive efforts, with a sudden increase in drive axle speed.
Splice	Joint between the component ends at the building machine. Beveled splice has a 45-degree angle. Butt splice angle is 90 degrees. Tapered splices range from 8 degrees to 30 degrees. Overlap splices extend up to 5 millimeters over each other.
Spring rate	Rate at which spring force grows as deflection increases, due to an applied load. Traditionally, it has been expressed in pounds per inch and is also referred to as spring stiffness. Measurement is conducted at rated inflation pressure.
Sprung weight	Total weight supported by the suspension.
Staggered tread blocks	Tread blocks of varied sizes and pitches to reduce vibration, reduce noise, and improve steering precision.
Standard radial	Tire aspect ratio of 80 and above.
Static loaded radius	Distance from the center of the wheel to the pavement, at rated load and inflation under static conditions.
Stitcher	Building machine roller used to apply pressure on a tread or other component to remove trapped air and improve component-to-component adhesion.
Tag axle	Free rolling axle located behind the last driven axle. Allows increase in GVW allowance.
Tandem axle	Two or more axles, one set behind the other.
Tenacity	Cord strength, frequently expressed in g per denier.
Toe-in	Alignment setting where tires on the same axle are closer together at the front than at the back, when measured at the spindle height.
TSR	Technically Specified Rubber.
Ton-mile	Movement of one ton of freight one mile. Ton-miles are calculated by multiplying the weight of the shipment by the distance hauled. Ton-miles per hour (TMPH) allow a productivity determination by including time. TMPH (and TKPH, ton-kilometers per hour) ratings are used for off-road earthmover tires.

Traction	Grip and friction between the tire and pavement. Tractive force is the corresponding force developed during braking or acceleration.
Tread	The tread is the component of the tire that is in contact with the road and provides wear resistance, good traction characteristics, fuel economy and service-related damage.
Tread base	Component underneath the tread that is designed to ensure good adhesion between the tread and the tire casing, and to dissipate heat from the tread. The shoulder region, which, in larger tires, is co-extruded as a unique component, also assists in heat dissipation.
Tread chimney	Term used to describe a compound in the tread to discharge static electricity, which builds up on the tire, particularly with silica tread compounds.
Tread pattern	Design of the tread through arrangements of voids, ribs, slots, lugs, sipes, and grooves.
Turn-up	The portion of the body ply that wraps under the bead and extends up the sidewall to the ply ending.
Twist	Number of turns per unit length in a cord or yarn; direction of twist.
Understeer	Tendency for front axle to lose grip before the rear axle grip is lost.
Under-tread	Compound between the bottom of the tread component and the cushion or top belts. Sometimes referred to as the base.
Uniformity	Measure of the tire's ability to run smoothly and vibration free; sometimes measured as tire balance, radial force variation, or lateral force variation.
Variation	It is the change in radial force as the loaded tire is rotated. Radial force variation will cause the vehicle to give a rough ride (as if on a poorly surfaced road). It may cause the tire to wobble and is due to irregular tire component dimensions. Lateral force variation is a summation of the lateral 1st, 2nd, 3rd, etc., harmonics.
Voids	Condition where compounded rubber does not flow sufficiently to fill the mold tread design, sidewall, etc., leaving areas unfilled areas. Sometimes referred to as lights and may be caused by component profile shape.
Warp	Cords in a tire fabric that run lengthwise.
Wheelbase	Distance between the front and rear axles.
Weft	Cords in a fabric running crosswise.
Yarn	Assembly of filaments can be either clockwise ("S" twist) or counterclockwise ("Z" twist); twist imparts durability and fatigue resistance to the cord, though tensile strength can be reduced.
Zipper break	Circumferential rupture of a sidewall's body ply cords or reinforcements.

1 Introduction to Tire Engineering

1.1 INTRODUCTION TO THE GLOBAL TIRE INDUSTRY

The global tire industry has shown remarkable growth, reflecting the growing global economy and the corresponding increase in standards of living. Discounting the impact of the pandemic due to COVID-19, in the year 2020, total global tire production had been expected to exceed three billion units. Though there is considerable variation in researcher's estimates for both previous and future production, regardless of the source, the growth of the tire industry has been largely linear over time, with an anticipated minor dip in 2021 and 2022, and then a return to historic linear growth rates from 2023 onwards (1–3). Other than a dip in 2009, the raw materials supply base for the industry has shown similar growth rates and is expected to exceed $110 billion by 2024, illustrating the impact of the industry on global economic growth (4).

Geographically, most of the future industry growth will be in East and South-East Asia, and will be driven by a number of important factors, first of which is a strong desire for car ownership, despite car sharing and other inner-city demographic trends. Secondly, increasing transportation growth and efficiency will continue to drive the trucking industry and associated truck and bus tire demands. Thirdly, national economic policies will influence future investments in new capacity. Tire manufacturing is labor intensive, capital intensive, and requires a high level of skilled labor, thereby supporting governmental policies on employment. This point will be reviewed in Chapter 8, on tire manufacturing.

With regard to industry growth estimates, there is typically a high degree of variability in the industry reports (5, 6). This is due to many factors, such as the quality of the raw data being collected, what tires to include, and the proprietary nature of data tire companies are willing to release. This discussion on tire engineering and technology will focus on only the main line tires, which cover products for the consumer and commercial markets, i.e., racing, passenger car, light truck, commercial truck and bus, farm, heavy-duty off-road, and aircraft tires. Though much of the technology still applies, bicycle, motorcycle, and light utility applications are largely excluded (Figure 1.1).

Within the industry, there have been two core business trends, firstly, the merging or acquisition of manufacturing companies in North America and Europe, driven by the need for economy of scale, and secondly, the emergence of new companies in Asia. In China, the government has encouraged the merging of companies to achieve economy of scale, to improve quality, and to protect employment. For example, tire factories require a certain throughput rate to be economic, in the order of 20,000 to 25,000 passenger tires per day or 270 tons of products, depending on the tire types.

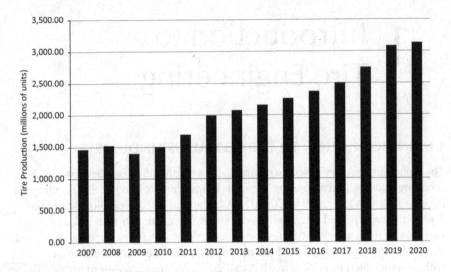

FIGURE 1.1 Global Tire Production (1)

The merging of smaller companies helps to facilitate more economic manufacturing output rates. Otherwise, manufacturing costs can undergo a rapid exponential increase as output drops, which has been one of the underlying reasons for the closing of many plants in traditional markets. Furthermore, the global market has evolved into essentially four regions, defined by the level of economic development and the quality of the transportation infrastructure. This is reflected in the number of tire manufacturing companies and the location of their headquarters (Table 1.1).

TABLE 1.1
Global Tire Manufacturing Participation and Revenues (7, 8)

America	Europe	South Asia	North Asia	Company	Revenue (billions $)
Goodyear Tire	Continental	Apollo	Bridgestone	Bridgestone	27.22
Cooper Tire	Michelin	Ceat	Cheng Shin	Michelin	25.40
Titan	Nokian	JK Tyre	Hankook	Goodyear	15.37
	Pirelli	MRF	Kumho	Continental	13.11
		GiTi	Linglong	Sumitomo	6.81
			Nexen	Pirelli	6.15
			Sailun	Hankook	6.12
			Sumitomo	Zhongce	4.80
			Toyo	Yokohama	4.14
			Yokohama	Cheng Shin	3.74
			Zhongce	MRF	3.50
				Toyo	2.95
				Cooper	2.85
				Apollo	2.17

TABLE 1.2

Tire Demographic Trends and General Focus for Different Regions (5)

Americas	Europe	North Asia	South Asia
Uniformity	Traction	Traction	Load capacity
Fuel economy	Noise	Wear	Wear
Wear (long haul)	Rolling resistance	Reliability	Durability
Safety	High speed	Rolling resistance	Economy

The emergence of four distinct regions or markets is driven by consumer requirements, infrastructure, and the local or national regulatory environments. What is common between the regions is a similar manufacturing base, i.e., the same type of production equipment and the same global raw materials supply base. The differences are with respect to product design, construction, and the compound line-up of the materials to meet the market needs. Table 1.2 is a simplified snapshot of regional performance parameters that regulators and consumers consider in tire performance. Though parameters such as traction and vehicle stopping distances are important in all markets, some performance parameters will tend to receive more emphasis than others, depending on the market.

Given the nature of the various markets, there are still many factors in tire technology in common, which are applicable to all markets, and this will now be covered. Five aspects of tire engineering and technology will be discussed: tire design, construction, materials technology, manufacturing, and end-product performance. There are many subsets to each of these five core areas. For example, in manufacturing, quality control and tire uniformity are of the utmost importance. Each tire manufacturer will have its proprietary procedures and processes to meet their product design requirements. This review will therefore primarily focus on only high-level variables, upon which practical manufacturing and quality assurance procedures could be developed.

1.2 TIRE TECHNOLOGY

The modern radial tire consists of a wide variety of chemicals, fabrics, and types of steel, rendering it the most complex composite in mass production (9, 10). This is due to the complexity of the material blends or compounds used in the tire, the number and different types of compound formulations, the complexity of fabric and steel reinforcements, and then the high level of uniformity and consistency necessary for the tire to perform on today's vehicles. In terms of both volume of production and consumer awareness, tires fall into essentially seven categories, which are based on vehicle application. There are tires for racing vehicles, passenger vehicles, and light trucks, where gross vehicle weights typically can reach up to 7250 kg. In such tires, significant quantities of fabric are used as reinforcement. Larger tires, such as those for heavy trucks and off-road vehicles tend to use much more steel wire reinforcement, whereas farm and agricultural vehicles, and large aircraft tend to use fabric reinforcements. In addition, there are a range of specialty tires which include those used on motorcycles, solid fork-lift trucks, light construction equipment, and golf cars. Regardless of the design or application of the tire, all tires must fulfill a fundamental set of functions:

1. Provide load-carrying capacity
2. Provide cushioning and dampening
3. Transmit driving and braking torque
4. Provide cornering force
5. Provide dimensional stability
6. Resist abrasion
7. Generate steering response
8. Exhibit low rolling resistance
9. Provide minimum noise and minimum vibration.

Three factors govern a tire's design and construction: i) the vehicle type and the mission profile for which the tire has been designed; ii) performance parameters, such as endurance and wear; and iii) handling characteristics, such as minimum vibration, high driver comfort, and steering precision.

The performance envelope of a tire describes the tire's characteristics in response to the application of load, torque, and steering input, resulting in the generation of external forces and deflection. All of the mechanical properties are interrelated, such that a design parameter affecting one performance factor will influence other parameters, either positively or negatively (8, 9). The result is a complex set of forces acting on a rolling tire on a vehicle (Figure 1.2).

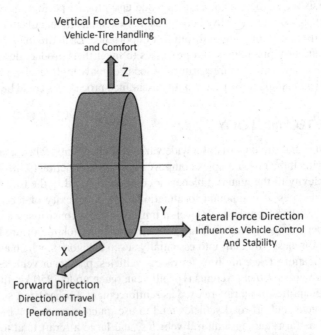

FIGURE 1.2 Tire Functions and Primary Forces Acting on a Mounted Tire

The tire–axis system is the center of the tire–road surface contact, as shown in Figure 1.2. The X axis is the intersection of the wheel plane with the road and is considered to be positive when travelling forward. The Z axis is perpendicular to the road plane and the downward force or load is considered positive, with the corresponding upward force being negative. The Y axis is in the road plane and forces in this plane can cause an overturning moment and resistance to rotation or rolling resistance. This representation of the tire axis system depicts the vehicle axis system of an automobile, when the tire is in a front steering position (9). When a vehicle is turning, the steering axle tire has a self-aligning torque or force and a lateral force vector. The forces acting on a tire can thus be broken down into three fundamental vectors: the vertical forces control the vehicle esthetics and comfort, the lateral forces impact on vehicle control, and the longitudinal or forward forces control performance, such as rolling resistance (Figure 1.3).

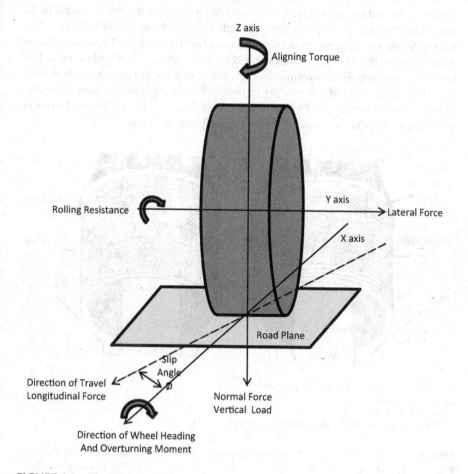

FIGURE 1.3 Tire Forces and Moments

1.3 TIRE CONSTRUCTION

The modern tire found in today's global market falls into one of two types, either a bias ply construction or a radial. There have been other construction derivatives, such as when belts were applied to bias tire constructions, sometimes referred to as belted-bias constructions, though none were ultimately successful.

In bias tires, the fabric plies are applied in a diagonal configuration extending from bead to bead, and the fabric cords lying at an angle of between 20 to 40 degrees to the centerline. Each successive layer of reinforcing fabric is applied in the opposite direction to the previous layer resulting in a composite network of layers running in opposing directions (Figure 1.4). The plies are anchored by bead cables. In larger tires, there can be multiple beads, depending on the number of plies used in the construction of the tire. The bias tire construction has been in production since the beginning of the tire industry and is still used today in many parts of the world, primarily for heavy trucks in on-/off-road applications and on underdeveloped road systems, farm tires (front tractor and implements for flotation), and aircraft tires (lightweight, and the prohibitive costs associated with radial tire certification). In the case of heavy trucks, with gross vehicle weights (GVW) from 10 tons to 40 tons, bias tires can still out-perform modern radial constructions under off-road conditions, due to their ability to roll over rocks or other rough surfaces. The load rating of bias truck tires is nominally determined by the number of plies. For example, a bias truck tire with a load rating of 'G' or a ply rating of "14" would typically have the corresponding number of layers of nylon or polyester fabric, i.e., 14, and be rated to carry a defined load at the tire's designed optimum inflation pressure.

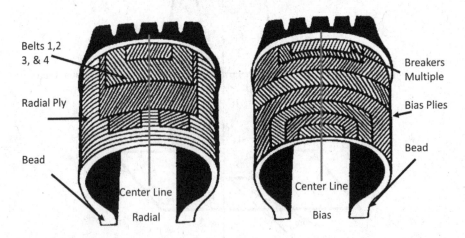

1. Bias tire plies made from nylon or polyester.
2. Radial tire ply is steel.
3. Bias truck, farm, and aircraft tire market still strong.
4. Bias passenger tire market still present in some Asian regions.

FIGURE 1.4 Bias and Radial Tire Constructions (9, 10). The first belt of a radial tire, Belt #1, is the first above the ply. It can be full width, or for large truck tires sometimes split in two parts.

The pneumatic radial tire was first developed in the 1950's, successfully commercialized by the early 1970's, and has now largely replaced the bias construction in those markets with fully developed road networks. Radial tires have a much less rigid casing construction, where flexible ply cords extend transversely or radially from one bead to the other bead, the bead again serving as anchors for the individual cords. The cords are at a 90-degree angle to the centerline of the tire. The ply cords in passenger tire tires are typically polyester, whereas, for heavy-duty truck tires, steel wire, designed for good fatigue resistance, is used. Below the tread and above the ply are steel wire belts made into a rigid composite hoop, which deflects the ply cord or ply-line but provides a firm foundation for the tread, creating a flat footprint and thus allowing the radial tire construction to out-perform the bias construction in terms of wear performance, traction, rolling resistance, and tire–vehicle handing (Figure 1.5).

Many parts of the world still use significant numbers of bias tires but the underlying trend in tire design is radialization. The European, Japan, and North American markets are essentially 100% radialized, with central and southern Asia still using a large number of bias tires, particularly on heavy-duty trucks. Compared to bias tire constructions, radial tires, such as the truck tire illustrated in Figure 1.6, offer the following advantages:

1. Improved tread wear due to less squirm or lateral scrubbing, as a result of the flat steel belts below the tread
2. Better traction due to the flatter footprint and more flexible sidewalls
3. Lower rolling resistance
4. Less noise generation
5. Retreadability
6. Higher load-carrying capability (better efficiency)
7. Lower operating temperatures in service, indirectly contributing to lower rolling resistance and better vehicle fuel consumption
8. Flexible casing, leading to improved damping characteristics and, in turn, improved driver comfort.

Unlike radial ply cord wires, the belt cords are rigid, with low extensibility. The number of belts will depend on the type of tire, with automobile passenger car tires

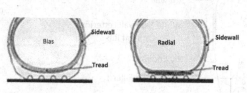

Bias construction, on turning tread and sidewall shift together

Radial construction, tread footprint is stable, while sidewall will flex

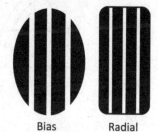

Bias   Radial

Radial tire showing improved footprint stability

FIGURE 1.5 Bias *Versus* Radial Construction (5)

FIGURE 1.6   Cross Section of a Radial Truck Tire (10)

(PCR) typically having two, heavy truck tires (TBR) having four, and large off-road or earthmover tires (OTR) with six layers. Figure 1.7 is a simplified illustration, showing the components of a radial tire used on vehicles. The major components of the tire can be described as follows:

> *Tread:* The tread is the component of the tire that is in contact with the road and provides wear resistance, good traction characteristics, high fuel economy, and has good damage resistance.

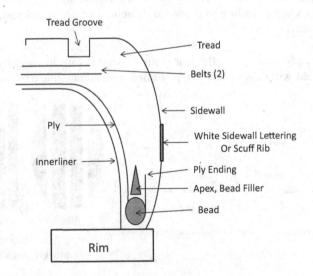

FIGURE 1.7   Major Components of an Automobile Tire

Tread Base: The component underneath the tread that is designed to ensure good adhesion between the tread and the tire casing and to dissipate heat from the tread. The shoulder region, which, in larger tires, is co-extruded as a unique component, also assists in heat dissipation.

Tread Chimney: Term used to describe a compound in the tread designed to discharge static electricity which builds up on the tire, particularly with silica-based tread compounds

Cushion: The rubber layer between the tread base and the tire belt assembly

Sidewall: Provides long-term weathering protection and casing durability. It also protects the tire from impact and curb-scuffing. It is compounded to resist fatigue and flex-related cracking.

Bead Region: The bead or lower region of the tire consists of the following components: 1) a steel wire bead ring, which anchors the ply and also locks the tire onto the wheel or rim; 2) the apex or bead filler; 3) a fabric insulation layer, sometimes called a "flipper", to prevent abrasion of the chipper and bead wire; 4) chipper wire, which protects the ply from rim damage; and 5) the chafer, which is the abrasion-resistant rubber component in contact with the rim. The apex or bead filler compound is formulated for dynamic stiffness, to enable stress distribution in the lower sidewall region, and provide good tire-vehicle handling properties. The bead insulation compound must have good steel bead wire adhesion and also good ply compound adhesion. The chafer, toe guard, or rim strip compound protect the plies from rim abrasion and help seal the tire to the rim.

Plies: Ply cords consist of steel or fabric, extend from the bead on one side of the tire to the opposite side, and serve as the primary reinforcing member of the tire. They provide the strength of the tire that enables it to retain air at high pressures. Automobile radial tires can have one or two layers of polyester cords, whereas bias truck tires can have many more, necessitating the use of multiple bead cables.

Belts: Layers of steel wire that form a hoop under the tread. They restrict deformation of the casing, provide rigidity to the tread region, and provide a stable foundation for the tread components. This allows improved wear performance, vehicle stability, handling performance, damage resistance, and protection of the ply cords.

Overlay: Calendered fabric layer, applied alternatively as strips, above the belts and under the tread, to help the belts withstand centrifugal forces as the tire rotates at high speed

Shoulder Wedge: A stiff elastomeric component located at the edge of the belt assembly and designed to provide stability to the shoulder region of the tire, provide a uniform footprint, and to help dissipate heat.

Inner Liner: A membrane consisting of compounded low-permeability rubber, such as bromobutyl rubber (BIIR) or chlorobutyl rubber (CIIR), the function of which is to retain compressed air inside the tire. The liner typically reaches from the bead on one side of the tire to the opposite side, thereby providing a seal.

Gum Strips, Barrier, Squeegee: Strip of calendared rubber used to cover belt
or ply endings, provide an interface between the inner liner and ply, and
promote component-to-component adhesion. Many times the terms gum
strips, barrier, and squeegee are used interchangeably.

Up to 14 different compounds can be found in a truck tire, and six different types of
steel wire cord and fabric reinforcement.

1.4 MISSION PROFILE AND DESIGN ENVELOPES

There are eight major types of pneumatic tires

1. Racing
2. Two-wheeler applications, such as tires for motorcycles and bicycles
3. Passenger car tires (PCR)
4. Light truck (radial light truck, RLT)
5. Medium or heavy truck, sometimes referred to as truck and bus radials
   (TBR)
6. Farm
7. Off-road (OTR)
8. Aircraft

There would also be sub-categories for each tire type, depending on the mission profile. For example, there are several types of automobile tires, such as high-mileage,
all-season, base-line and broad-market tires, traction, and high-performance tires.
Each type of tire will have a specific type of tread pattern and tread compound,
designed to meet the design targets. Though there are many exceptions, Table 1.3

TABLE 1.3
Nominal Tread Pattern Designs (9)

Tire Category	Tread Pattern Design	Application	Net-to- Gross Tread Rubber Contact Area	Performance Target
1	Centerline solid rib	High mileage	High	Tread wear
	Outer rib-block configuration			Low rolling resistance
2	Block-rib	All-season	Medium high	Tread wear
		Broad market		Traction
		Baseline		Low rolling resistance
3	All-block	Traction	Medium low	Traction
4	Central groove	Traction	Medium low	High traction
	Outboard blocks	High performance		
5	Directional	High performance	Low medium	High traction, handling
	Asymmetric			

and Figure 2.11 provide a general summary of the type of tread pattern to be initially selected when designing a new tire for a given mission profile, such as a base-line, high-performance, or high-mileage passenger tire.

Light truck tires (RLT), such as those used on pick-up trucks, tend to follow the same design principles as those for automobiles. The ply is polyester, and the tire has two steel cord belts. The primary difference is the tensile strength or tenacity of the polyester cords, which is necessary for the heavier vehicle weights; depending on the load-carrying capacity of the tire, two fabric plies may be used.

Like automobile and light truck tires, medium- and heavy-duty truck tires are also classified by the mission profile or service application. Trucks have a classification system defined by the gross vehicle weight. There are eight classes (Table 1.4) and, depending on the loads and type of service, they cover the range of applications from light service city pick-up and delivery vehicles through to heavy-duty long-haul rigs.

Truck tire design, compounds, and construction have been optimized for each of the vehicle classes and axle positions, but essentially fall into one of five types, namely steer axle tire, drive axle tire, trailer axle, on-/off-road steer axle tire, and on-/off-road drive axle tire. Again, there are many subsets to this, such as specific tires for buses, recreation vehicles, straight trucks, and tractor–trailer combinations. In all cases, the tread patterns are essentially a rib design, a lug pattern, or a combination of the two, depending on the design profile for the tire (Figure 1.8). Similarly, farm tires are designed for the wheel position, the tractor front or steering axle, which tend to be primarily bias in construction, with a rib tread design, rear drive axles, which largely use radials with a lug tread design, and implements which are bias and tend to have flotation qualities, i.e., which will have a wide footprint to avoid sinking into loose soils and to minimize soil compaction. Earthmover tires

TABLE 1.4
Society of Automotive Engineers (SAE) Truck Classification (11)

Class	Gross Vehicle Weight		Tire Casing Construction	Some Vehicle Type	Some Applications
	Imperial Pounds (lb)	Maximum Kilograms (kg)			
1	6000 or less	2722	Fabric ply	Pick-up truck, bus	City service
2	6,001–10,00	4536	Fabric ply	Pick-up truck, bus	City service
3	10,000–14,000	6350	Fabric ply	Pick-up truck, bus	City service
4	14,001–16,000	7258	Fabric ply	Delivery van	City service
5	16,001–19,500	8845	Steel ply	Delivery van, light bus	City service
6	19,501–26,000	11,794	Steel ply	Bus, truck	City, metro
7	26,001–33,000	14,469	Steel ply	Bus, straight truck	Metro, regional
8	33,001 and over	14,970	Steel ply	Line haul, construction, Coaches	Interstate, heavy duty

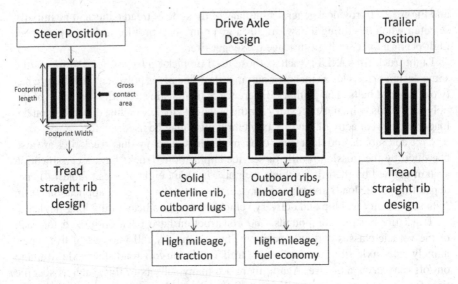

FIGURE 1.8   Truck Tire Tread Design Parameters

would typically fall into one of six categories, depending on the application, which are backhoe, loader, truck, grader, scraper, and dozer. Aircraft tires are designed for either commercial aircraft, business and general utility applications, helicopters, and the military. Tires for each of these applications will have a unique construction, reinforcement, and an optimized set of compounds designed to ensure that the tire meets the required service demands.

1.5   DIMENSIONS AND NOMENCLATURE

The terminology used to describe tire and rim dimensions is explained in Figure 1.9 and is commonly used throughout the tire industry to describe size, growth, and

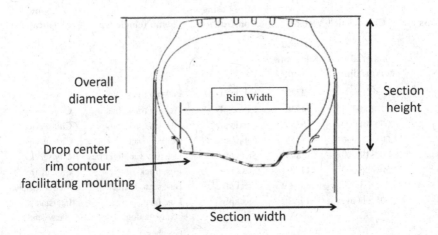

FIGURE 1.9   Dimensions of a Tire Mounted on a 15 Degree Rim (10)

wheel well clearance factors, in addition to computation of variables such as load capacity and revolutions per unit distance traveled. The section width is the width of the tire from sidewall to sidewall. The section height is the distance from the bottom of the bead region of the tire to the top of the tread. These two parameters are important in describing the tire size seen on the sidewall.

Tire constructions being manufactured today can be described as falling within one of three groups: bias tires typically requiring a tube to help maintain inflation pressure, tube-type radial tires which also require a tube, and tubeless radials which use a chlorobutyl or bromobutyl impermeable inner liner. Three basic tire size descriptions are thus:

1. Conventional size tires used on flat base rims, normally for tube-type tires
2. Conventional size tires used on drop center rims for tubeless tires. Such tires typically have a nominal aspect ratio of 80 (the section height is 80% of the section width)
3. Metric sizes or low-profile sizes mounted on drop center rims for tubeless tires and where the aspect ratio is lowered to 75, or to as low as 25 for ultra-high-performance tires.

There are a number of additional terms illustrated in Figures 1.9 and 1.10 that are used in describing tire dimensions, and which can be summarized as follows:

Overall Diameter: Mounted and inflated, but not loaded, new tire diameter;
Section Width: New tire section width, excluding scuff ribs and lettering;
Section Height: Distance from the tire centerline to the rim;
Aspect Ratio: Section height divided by section width, expressed as a whole number. A conventional-sized tire has a nominal aspect ratio of 80.
Static Loaded Radius: Distance from the road surface to the center of the rim;
Minimum Dual Spacing: The minimum required distance between the centers of two mounted rims to prevent the adjacent tire sidewalls from abrading against each other;
Loaded Section Width: Width of the loaded tire section, typically above the road contact area;
Footprint Length: Length of the loaded footprint;
Footprint Width: Footprint width;
Gross Contact Area: Total footprint area;
Net Contact Area: Area of tread rubber compound in contact with the pavement, excluding voids;
Rim Width: Distance between rim flanges (frequently defined in inches) and is specified for each size of tire. For example, a 11R22.5 tire would use a 8.25" rim;
Flange Height: Rim flange height above the rim base.

The tire size therefore includes the section width, aspect ratio, and rim size. In addition, letters can be added to describe the tire type:

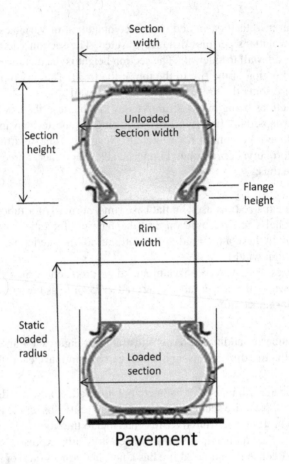

FIGURE 1.10   Illustration of Tire Dimensions Terms (10)

- R in the size description designates "radial,"
- D or (–) denotes "bias,"
- ML indicates that the tire is for mining and logging applications.
- P at the beginning of the size description denotes passenger automobile tires, and LT describes a tire for a light truck or large sports utility vehicle (SUV).

All of this information is displayed on the tire sidewall. By far the most popular type is low- profile or metric-sized tires. For a passenger car tire, a typical metric size such as a P255/45R16 would read as follow:

P   Passenger car
255   Section width in millimeters
45   Aspect ratio
R   Radial
16   Tire wheel diameter in inches

Conventional size tires are mostly found on truck and bus tires. A conventional truck tire, which might be used on buses, would be an 11R22.5, where:

11   Section width in inches
R   Radial
22.5 Rim diameter in inches

Such tires will have a nominal standard aspect ratio of 80. The sidewall will also show the load rating, load index, and a speed rating. Other sidewall markings include the following:

1. Safety warnings, primarily focusing on minimum and maximum inflation pressures;
2. Maximum inflation pressure:
3. Load rating, load index, and speed symbol (Table 1.5):
4. Speed rating, which may be included in the tire size description, e.g., P255/45ZR16 where "Z" defines the speed capability (Table 1.6):

TABLE 1.5
Tire Ply Rating (13, 14)

Load Range	Tire Ply Rating[1]
D	8
E	10
F	12
G	14
H	16
J	18

[1] Rating historically related to the number of fabric plies in a bias tire

TABLE 1.6
European Load Index Numerical Code for Tire Load-Carrying Capability

Load Index	Load-Carrying Capability (kg)
144	2800
145	2900
146	3000
149	3250
152	3550
155	3875
160	4500

5. Manufacturer's name, logo, and other trademarks:
6. Construction information such as the number of plies and the ply material;
7. Uniform tire quality grade (UTQG) information which provides ratings on the tire's tread wear potential, traction capability, and operating temperature:
8. Department of Transportation (DOT) serial number. The tire DOT number is an alphanumeric code on the sidewall of every tire and provides the following information:
    a. The sidewall stamp begins with "DOT" which confirms that the tire meets the United States Department of Transportation safety standards;
    b. The first two alphanumeric characters assigned by the United States Department of Transportation define the tire manufacturing plant;
    c. The next five characters are nominally assigned by the tire company and relate to the tire size and any manufacturing specifications
    d. The last four digits are, first, the week and then second, the year the tire was manufactured, e.g., 2118 which would be the 21st week of 2018. Prior to the year 2000, there were three numbers, e.g., 126 which would be the 12th week of 1996.

UTQG is a passenger tire rating system that illustrates a tire's potential performance for tread wear, durability, traction, and temperature resistance. UTQG ratings are required by the United States federal government for most types of passenger tires, and the UTQG ratings are molded on the tire's sidewall. The tread wear grade is a numeric rating, with a higher number suggesting longer tread wear capability. Most tires receive grades between 100 and 800. The traction grade is assigned on the basis of results from skid tests on wet pavements. Tires are graded AA, A, B, or C, with AA indicating superior wet traction. The temperature grade is assigned to tires tested at various speeds to determine the ability of a tire to dissipate heat. Tires are graded A, B, or C, with A indicating an ability to dissipate heat at higher speeds. The European Union has a similar standard for tire size descriptions, but, in addition, has a symbol for the country of manufacture (Figure 1.11).

The load-carrying capability of a tire is also dependent on the inflation pressure. Thus, for a given tire load range rating, the specific load-bearing capability will change with inflation pressure and the near-linear relationship is shown in Figure 1.12. Finally, as mentioned above, mounting and demounting of a radial tire is made possible because of a well in the center of the rim base, which is also referred to as the drop center. Rims for bias tires would have a detachable flange. One additional parameter influencing the maximum inflation pressure of a tire is the pressure rating of the rim. For example, many truck tire rims are designed for a maximum inflation pressure of 120 psi or 830 kPa. This is another important parameter in selecting tire and wheel systems for a vehicle.

1.6 TIRE SPEED AND LOAD RATING

Historically, the speed rating for a tire was included in the tire size description. For example it might read 205/65HR16, where the tire has a section width of 205 mm, aspect ratio of 60, is a radial tire, has a rim size of 16 inches, and "H" denotes the

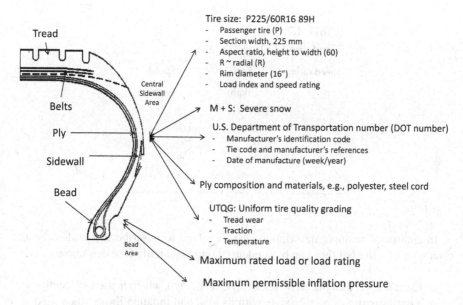

Tread

Belts

Ply

Sidewall

Bead

Central Sidewall Area

Bead Area

Tire size: P225/60R16 89H
- Passenger tire (P)
- Section width, 225 mm
- Aspect ratio, height to width (60)
- R ~ radial (R)
- Rim diameter (16")
- Load index and speed rating

M + S: Severe snow

U.S. Department of Transportation number (DOT number)
- Manufacturer's identification code
- Tie code and manufacturer's references
- Date of manufacture (week/year)

Ply composition and materials, e.g., polyester, steel cord

UTQG: Uniform tire quality grading
- Tread wear
- Traction
- Temperature

Maximum rated load or load rating

Maximum permissible inflation pressure

FIGURE 1.11 Tire Sidewall Information (12) https://en.wikipedia.org/wiki/Tire_code

speed rating, in this case 210 km/h or 130 mph (Table 1.7). The newer system has the speed rating and load index shown separately. In this case, the size would read 205/65R16 85H, where the index, 85, refers to load capability (1135 lb). The load index is an assigned number initially developed by the European standards organizations and which defines the load-carrying capacity of the tire. For passenger car tires, the load index is typically in the range from 70 to 105.

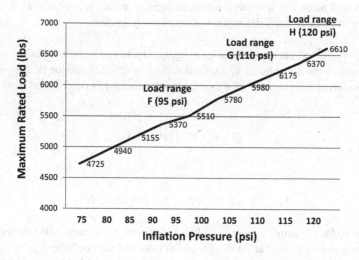

FIGURE 1.12 Maximum Rated Load With Increase in Inflation Pressure, Single Mounted 275/80R22.5 Truck Tire

TABLE 1.7
Tire Speed Rating (13, 14)

Speed rating	Speed (kph)	Speed (mph)
S	180	112
T	190	118
H	210	130
V	240	149
Z	270 and above	168 and above

In summary, multiple descriptions have evolved to describe the load-carrying capability of a tire but they can be condensed into essentially four definitions:

1. Load Range: designated by a letter (A, B, C, etc.), which is used to identify a highway tire of a given size, with its load and inflation limits when used in a given type of service;
2. Ply Rating: the term "ply rating" is a term which evolved from bias ply fabric reinforced tires, and is used to describe the tire load-carrying capability. The greater the number of bias fabric plies, the greater the load-carrying capability. For example, a tire with a ply rating of "7" would be equivalent to a load range "G" tire. The term "Ply Rating" is now a strength index unrelated to the number of plies in the tire.
3. Load Index: a numerical code defining the maximum load a tire can carry at a given speed, indicated by the speed symbol, and a given inflation pressure.
4. Symbol Mark (*): a symbol which defines an inflation pressure, load, and size of a tire under a defined set of service conditions.

The theoretical load-carrying capacity in kg, Q, of a tire at a given inflation pressure is a function of the air volume, V, nominal section width, SN, section height (H), and the nominal rim diameter Dr. The load-carrying capacity is determined from equation 1.1;

$$Q = K \times V^b \times (P + Po) \qquad (1.1)$$

where:

$$V = \text{tire volume} = \pi \times SN \times H \times (Dr + H) \qquad (1.2)$$

The term P, for pressure, is set at 240 kPa for 80 series tires and 250 kPa for lower aspect-ratio tires. Po is set at 70 kPa for radial tires and the coefficient, b, in the term V^b is set at 0.8. The coefficient, K, for tires of aspect ratios 20–55 is 1.60×10^{-6} and for tire aspect ratios of 60–80 is 1.58×10^{-6}.

In practice, the tire size is determined by rim dimensions, wheel well envelope, service load, service speed, and inflation pressure. For a truck tire, these factors are related by the equation

$$L = (6.075 \times 10^{-5}) \, K \times P^{0.7} \times S^{1.1} \, (D_r + S_d) \quad\quad\quad (1.3)$$

where L=load at 100 kph (kg); P=pressure (kPa); Sd=dimensional factor, section width adjusted for aspect ratio; Dr=rim diameter; and K=constant, dependent on vehicle speed. Using equations such as equation 1.1, developed by the Tire & Rim Association, the tire engineer can determine the optimum tire size for a specific application. Normally, load requirements are known so these equations can be used to calculate required service pressure. This process is then used in size and load-range selection. Tire industry standards are used extensively in the design process. Again, in practice, depending on the gross vehicle weight, most commercial truck tires, such as an 11R22.5 or 275/80R22.5, are designed to run optimally at an inflation pressure of 700 kPa or nominally 100 psi, whereas passenger car tires, such as a 205/55R16, are designed to run at 220 to 240 kPa (nominally, 32 to 34 psi). The potential load-carrying capacity is also a function of inflation pressure; the higher the inflation, the greater the load-carrying capability (Figure 1.12). The Tire and Rim Association has published tables showing inflation pressure versus load-carrying capacity for all tires and therefore provides design targets which the development engineer should meet. Table 1.8 illustrates a sample of such data, and more extensive information is shown in the Year Books published by the Tire and Rim Association, to which further reference is recommended (13).

1.7 TRENDS IN TIRE SIZES

There has been a significant increase in the number of tire sizes. This is due to a number of reasons: i) an increase in the number of vehicle platforms; ii) radialization; iii) introduction of low profile-size tires, being due to the advantages they allow

TABLE 1.8
Representative Sample of the Load and Inflation Pressure Table for Heavy-Duty Line-Haul Radial Truck and Bus Tires (13)

Rated load in lb	Size	Mounting	Inflation Pressure (in psi)								
			70	75	80	85	90	95	100	105	110
	11R22.5	Dual	4380	4580	4760	4950	5120	5300	5470	5630	5800
		Single	4530	4770	4990	5220	5430	5640	5840	6040	6240
	11R2.4	Dual	4660	4870	5070	5260	5450	5640	5820	6000	6170
		Single	4820	5070	5310	5550	5780	6000	6210	6430	6630
	275/80R22.5	Dual	4500	4690	4885	5070	5260	5440	5675	5795	6005
		Single	4500	4725	4945	5155	5370	5510	5780	5980	6175
	285/75R24.5	Dual	4540	4740	4930	5205	5310	5495	5675	5860	6175
		Single	4545	4770	4990	5210	5420	5675	5835	6040	6175

TABLE 1.9

PCR and RLT Tire Size Proliferation (13, 15, 16)

Dimension	Units	1951	1975	2003	2017
Number of sizes		21	91	254	462
Section width	mm	105–205	155–295	145–345	125–405
Aspect ratio		96–100	50, 60, 70, 80	30 to 80	25–85
Outside diameter	mm	545–845	560–765	540–840	145–840
Maximum rated load	lb	430–735	430–1000	370–1320	1653

in tire-vehicle handling and stability characteristics; and iv) an increase in vehicle weights (Table 1.9). Rim sizes have also increased as tire aspect ratios decreased, further allowing improvements in vehicle stability improvements and fuel economy.

1.8 SECONDARY FACTORS INFLUENCING TIRE DESIGN

In designing a tire for a new vehicle platform, several additional design parameters are of importance, specifically (17):

1. Static loaded radius
2. Rolling circumference
3. Tire spring rate

The static loaded radius is typically determined as part of the development cycle. This radius is controlled to a large degree by the tire pressure. However, preliminary estimates can be made using simple calculations, such as in equation 1.4:

$$\text{Static Loaded Radius} = Dr/2 + \left[k + \{ (Do - Dr)/2 - 17.5 \} \right] + 17.5 \qquad (1.4)$$

where Dr is the nominal rim diameter, k is 0.71 for tire pressures at 180 kPa and 0.76 at 240 kPa, Do is the overall diameter for the tire design and 17.5 is the rim flange height.

Similarly, the tire rolling circumference is sensitive to the load and inflation pressure. It is important in making estimates of a vehicle's potential fuel consumption correlations with velocity and distance data. Therefore, theoretical estimates can be made using equation 1.5:

$$CR = F \times D \qquad (1.5)$$

where CR is the theoretical rolling circumference, F is a coefficient, set at 3.05 for a vehicle at 80 kph (50 mph) and inflation pressure between 180 kPa and 240 kPa, and D is the overall design diameter.

Spring rate is defined as the change in spring force per unit increase in deflection. Spring rates typically have units of lbf/in. or N/mm, and an example of a linear spring

TABLE 1.10

Spring Rates for Truck Tires of Equivalent Outer Diameter (17)

Tire Size	Vertical Load on Tire and Rim	Force (lb/in)		
		Radial	Lateral	Tangential
295/75R22.5	5300	5550	1500	2600
11R22.5	5300	6450	1600	2950
10.00R20	5300	6650	1750	3250

rate is 500 lbf/in., i.e., for every inch the spring is compressed, it exerts an additional 500 lbf. Table 1.10 shows three tires, of similar outer diameters (41"), which can be substituted for each other without vehicle gearing or drive-train changes, but which show different properties.

Tire constructions will have an effect on the spring rate that, in turn, influences the suspension design or springs, shock absorbers, and air bag specifications. The suspension of a vehicle has five functions:

1. With the tire and wheel assembly, it supports the vehicle's weight
2. It transmits braking and driving forces to the tire
3. It cushions the ride
4. It secures the axle to the driveline
5. It maintains drive train alignment

The suspension and tires, therefore, function as a composite system to ensure adequate vehicle performance and driver comfort. The tire spring rate plays a critical role in achieving this interaction.

1.9 OFF-ROAD TIRES

Heavy-duty tires designed to operate in rough terrain, highway construction, and mining, or under unpaved road conditions, are termed "off-road tires" or OTR. They fall into one of six general vehicle applications:

1. Dozer
2. Scraper
3. Loader
4. Truck
5. Grader, or
6. Backhoe

Many tire companies further categorize off-road or earthmover tires as those which use rims of diameters of 25" and greater, whereas truck and bus tires have rim diameters of 24" and lower. For example, a 12.00R24 tire would be classed as a truck tire, whereas 14.00R25 up to 40.00R57 tires are OTR tires. Compound line-up and tire building systems also change. Such tires are rated for ton-mile-per-hour (*TMPH*) or ton-kilometers-per-hour (*TKPH*), which is analogous to the rolling resistance of a truck tire, in that it is largely a function of whole tire hysteresis, i.e., the higher the operating temperature of the tire, the lower will be its potential *TMPH* rating. The rating is calculated by multiplying the average load per tire by the average speed of the vehicle per hour;

$$TMPH = \left[\left(E + L \right) / 2 \right] \times \left[\left(M \times N \right) / H \right] \qquad (1.6)$$

where

E = empty load on the tire
L = loaded tire weight
M = round trip distance
N = number of trips
H = total hours of operation

The *TMPH* or *TKPH* rating is used in selecting tires for specific operations.

Most tire manufacturers, who produce large off-road tires, mandate nitrogen inflation rather than air. Because of the nature of the operating environment, vehicles have a greater potential for excessive braking, dragging brakes, and, in some instances, repair welding of rims with mounted tires, and, as a consequence, a vehicle fire can start, causing the inside of the tire to ignite and burn. To prevent auto ignition, nitrogen inflation is used. This also helps eliminate the risk of an explosion, due to excessive heat, which is more destructive than a tire blow-out or other rapid air loss condition. Other benefits include:

- Improved air pressure retention (IPR) or reduced air pressure loss rate (IPLR)
- Improved aging resistance due to less thermo-oxidative degradation of the tire compounds
- Reduced rim rust

1.10 FARM TIRES

Farm tires essentially fall into one of three categories: front steer axle, where, in many instances, they are bias ply; rear drive axle tires, which are typically radial and traditionally are larger than those on the front; and implement tires, which tend to be bias. Though the trend is toward radialization in all wheel positions, factors such as "flotation", reduced soil compaction, and stubble damage resistance (i.e., corn stalks after harvesting can be both ridged and sharp, and can thus cause tire damage) define the tires selected for different equipment and applications.

Farm tire types can be further subdivided. For example, rear radial tires can be developed for specific applications, such as narrow furrows, rice harvesting, cotton harvesting, backhoes and general-purpose equipment use, such as for wheat harvesting.

TABLE 1.11

Aircraft Tire Centrifugal Forces on Take-Off (18)

Take-off Speed (mph)	Centrifugal Force on Tread (lb)
100	4,000
200	16,600
300	38,500
400	67,500

1.11 AIRCRAFT TIRES

Tires used in aviation fall into one of four types, defined by the application: military, commercial, business jets, and rotary. The four primary development parameters around aircraft tires are i) high-speed capability, which specifically focuses on constraining centrifugal forces acting on the tire crown area during landing and take-off, ii) retreadability, improved by reducing whole tire hysteresis and heat generation, iii) the number of landings, and iv) radialization.

On aircraft take-off, as the tire leaves its footprint or deflection zone, it attempts to return to its normal diameter. However, due to centrifugal forces, the tread surface can show a bulge, briefly distorting the tire from its natural shape (Table 1.11). This sets up a traction wave in the tread surface; to minimize this, inflation pressure maintenance is of the utmost importance. Excessive traction waves can also cause tread groove cracking, leading to cut propagation under the tread.

1.12 TIRE MATERIAL COMPOSITION

Having noted the complexities in tire design and construction, and the compliance requirements, the materials are equally complex. Figure 1.13 breaks down the compound material composition into several classes, such as polymers, chemicals, and fillers such as carbon black. In addition, there are different types of steels and fabrics. The dominant material used in radial tires is natural rubber due to its high tensile strength, compounds with high tire component-to-component adhesion, and low hysteresis. Because of this, natural rubber technology will be discussed in more depth later (Chapter 6). Among the synthetic elastomers solution styrene butadiene rubber (SBR) consumption has increased due to the increasing use of silica to improve rolling resistance. New-generation functionalized solution polymers have further facilitated this increase over comparable polymers, such as emulsion SBR.

1.13 SUMMARY

Geometrically, a tire is a torus. Mechanically, a tire is a flexible, high-pressure container. Structurally, a tire is a high-performance composite built using elastomers, fibers, steel, and a range of organic and inorganic chemicals. Tire technology has thus evolved into a complex combination of science and engineering, that brings

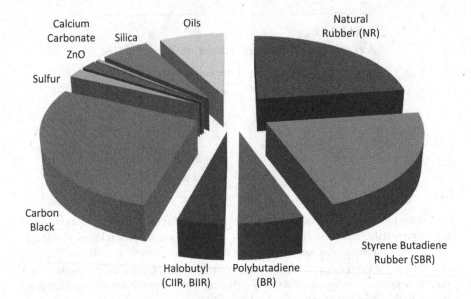

FIGURE 1.13   Typical Passenger Tire Composition (5, 19)

together a variety of disciplines. In the development of a tire, knowledge in the areas of tire geometry, dynamic tire behavior, the chemistry of component materials, and the technology of composite structures is essential. The result is a broad range of products designed to meet the needs of vehicle manufacturers and end-consumers for optimum performance under a wide variety of service conditions.

REFERENCES

1. United Nations Department of Economic and Social Affairs, Statistics Division. *Monthly Bulletin of Statistics*, Vol. LXXIII, No. 1181, p. 11. 2019.
2. Global Demand for Car and Light Commercial Vehicle tires. https://www.statista.com. 2019.
3. Mayer A. *The Future of Truck Tires to 2027*. Smithers Scientific Services, Akron, OH. 2017.
4. Market Focus. *Rubber World*, Vol. 262, No. 2. November 2019.
5. Rodgers MB. *Global Tire Trends*. ELL Technologies LLC, Cedar Park, TX. 2019.
6. Mayer A. *The Future of Global Tires to 2022*. Smithers Scientific Services, Akron, OH. 2017.
7. Global Tire Rankings 2019. *Rubber and Plastics News*.
8. Market Research Reports. www.marketresearchreports.com/worlds-15-largest-tire-manufacturers-revenue. 2019.
9. Rodgers B, Waddell W. Tire Engineering. In *Science and Technology of Rubber*. Eds. JE Mark, B Erman, CM Roland. Academic Press, Boston, MA. 2015, 653–696.
10. Rodgers B, D'Cruz B. Chapter 14 Tire Engineering. In *Rubber Compounding Chemistry and Applications*, Ed. Brendan Rodgers. CRC Press, Boca Raton, FL. 2015.
11. Fitch JW. *Motor Truck Engineering Handbook*. SAE, Warrendale, PA. 1994.
12. Tire Sidewall Labelling. 2020. https://en.wikipedia.org/wiki/Tire_code.

13. *Tire and Rim Association Year Book*. Copley, OH. 2016.
14. *Engineering Design Information*. The Tire and Rim Association Inc. Copley, OH. 2005.
15. Clark SK. *Mechanics of Pneumatic Tires*. U.S. Department of Transportation, National Highway Safety Administration. 1975.
16. Lindenmuth BE. An Overview of Tire Technology. In *Pneumatic Tire*, Eds. AN Gent, JD Walter. National Highway Traffic and Highway Administration (NHTSA), Washington, DC. 2005, 3–37.
17. Ford T, Charles F. *Heavy Duty Truck Tire Engineering*. Society of Automotive Engineers, SP-729, Warrendale, PA. 1988.
18. *Aircraft Tire Care and Maintenance*. The Goodyear Tire & Rubber Company, Akron, OH. 2017. https://www.goodyearaviation.com/resources/pdf/aviation_tire_care__2017.pdf.
19. Rodgers B, Waddell W. The Science of Rubber Compounding. In *Science and Technology of Rubber*. Eds. JE Mark, B Erman, CM Roland. Academic Press, Boston, MA. 2015.

2 Tire Tread Technology

2.1 INTRODUCTION

Though every component in a tire plays a role in ensuring that the product meets its required design envelope and mission profile, the tread composition and design are of considerable importance. The tread is also the part of the tire that attracts the most research and development efforts, due to its impact on tire rolling resistance and ultimate tire life. Furthermore, this part of the tire has the greatest impact on wear performance, traction, vehicle stopping distance, whole-tire hysteresis, which relates to fuel consumption, damage resistance, tire operating temperature, and tire-vehicle handling characteristics.

The tread is a thick extruded profile that surrounds the tire casing. It typically consists of up to three components, though, in many instances, such as is seen in high-performance tire treads, up to six different compounds can be produced, using a pentaplex extruder with five barrels and a cushion applicator. The three primary components are:

1. The tread, which is in contact with the pavement of the road surface,
2. The base or under-tread, which is a transition component adhering the tread to the tire casing while also providing a means to lower tread operating temperatures and improve rolling resistance, and
3. The cushion, typically up to 2 mm in gauge and providing additional adhesion to the belt compound. When used, it is typically applied at a cushion mill or small extruder downstream along the tread extrusion line.

Depending on the tire design, a wing, or "mini-wing", produced from sidewall compound, can be added to the edges of the tread. In many cases, such a construction is termed a "sidewall-over-tread" and, though the building of such constructions is an additional step in manufacturing complexity, it can offer end-product performance benefits (Figure 2.1).

In many respects, the tread performance is determined by four design parameters, namely i) the tread compound, ii) the tread pattern, iii) the footprint shape and pressures, and iv) the tread radius, which affects the footprint pressure in the tread shoulder area. Environmental conditions also have an effect on tire tread wear. Such external variables include the vehicle type and drive train design, speeds and loads, driver characteristics, weather and topography, and road surface conditions. Focusing on the tire tread, each of the four design parameters will be considered.

2.2 TREAD COMPOUNDS

As is the case for all tire compounds, the tread compound formulation consists of essentially five types of materials: the polymers or elastomers, reinforcement materials or fillers, antioxidants and other compound protectants, processing aids, and the

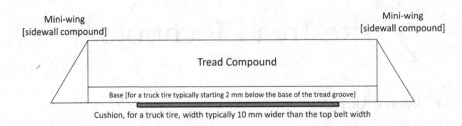

FIGURE 2.1   Extruded Tread Profile Illustrating the Tread, Base, Cushion, and Mini-wings

vulcanization system (1, 2, 3). The selection of the elastomers will depend on the mission profile and design parameters for the tire but will, in all cases, be selected to provide wear resistance, traction under both wet and dry conditions, low rolling resistance, and contribute toward good vehicle fuel economy, and resistance to damage caused by cutting or tearing of the tread. Tread compound development is a complex field in materials science, polymer chemistry and technology, and chemical blending to optimize the required mechanical properties. For example, as a general guideline, hard tread compounds have long wear characteristics but poor traction, whereas soft compounds have good traction but poor wear characteristics. Rolling resistance, damage resistance, and contribution to tire-vehicle handling qualities further impact formulation composition. The types of materials used in a tread formulation can be further described as follows;

1. Natural rubber, or polyisoprene, is the major elastomer used in radial tires due to its high resilience and low heat-buildup properties (low hysteresis), high tensile strength, and high tear strength
2. Styrene-butadiene co-polymer (SBR) is a synthetic rubber used in tread compounds or often substituted in part for natural rubber. Due to its higher glass transition temperature (Tg) and its effect on final compound Tg, it can allow improvements in tire traction and handling performance. Figure 2.2 shows typical tire tread applications for various polymers. There are two types of SBR, depending on the production process, i.e., solution and emulsion SBR
3. Polybutadiene is used in combination with other rubbers because of its low hysteresis, high abrasion resistance, and improved fatigue resistance
4. Halobutyl rubber, such as bromobutyl, is used in tubeless tire inner liner compounds, because of its low air permeability. However, it has also been used in tread compounds, with its high hysteresis but graduated glass transition temperature (Tg), enabling significant improvements in traction with little trade-off in rolling resistance
5. Carbon black makes a high percentage weight contribution to the rubber compound and has three functions, improving compound reinforcement and abrasion resistance, improving polymer and compound processability during manufacture, and allowing final compound cost optimization
6. When used with a silane coupling agent, silica, together with carbon black, allows reduction in heat build-up, lower operating temperature, improved rolling resistance, and improved traction performance. Though the new

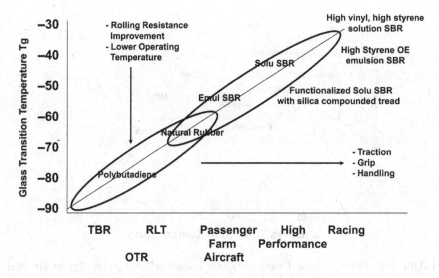

FIGURE 2.2 Tread Compound Polymers by Tire Application (4)

performance-grade silicas, termed highly dispersible silica, or HDS, were initially developed for low rolling resistance and superior traction passenger tires, they are now also found in light truck radial tires and, more recently, in commercial heavy-duty truck and bus tires

7. Sulfur crosslinks the rubber molecules in the vulcanization process
8. Vulcanizing accelerators are complex organic compounds that speed up the vulcanization process
9. Activators help initiate the vulcanization process. The main activator is zinc oxide, which combines with stearic acid to form zinc stearate, which, in turn, serves to activate the accelerators (1)
10. Antioxidants and antiozonants prevent compound oxidation and anaerobic degradation, and reduce tread sidewall cracking due to the action of sunlight and ozone.

The most fundamental property of a tread compound is the glass transition temperature, which, in turn, can affect the dynamic moduli, E' (in tension) and G' (in shear), the corresponding loss moduli and then, in turn, the hysteresis of the compound. As an empirical guide, the larger the tire line, the lower will be the tolerated heat build-up and hysteresis, and therefore the lower will be the required typical compound Tg. Similarly, Tg has near-linear relationships with both wet traction (Figure 2.3) and tire rolling resistance (Figure 2.4), whereas a more exponential relationship with tread wear is observed (Figure 2.5).

Such observations have allowed the development of a series of empirical relationships for tread compounds (1):

1. There is an improvement in wear resistance as the compound Tg decreases
2. With regard to polymer microstructure, which describes the arrangement of monomers in a polymer, wet grip or traction improves in a near-linear manner with increases in compound Tg

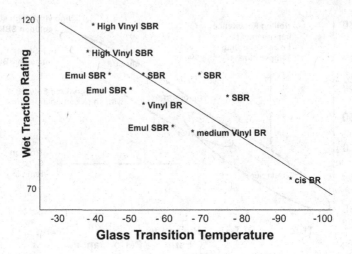

FIGURE 2.3 Effect of Tread Compound Glass Transition Temperature (Tg) on Tire Wet Traction (4)

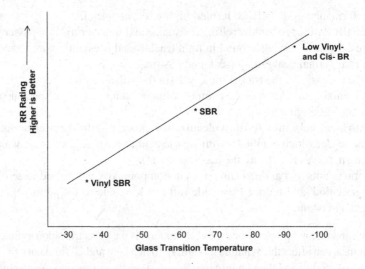

FIGURE 2.4 Tread Compound Tg and Tire Rolling Resistance (4)

3. Increase in 1,2-vinyl butadiene content (Figure 7.1) in polybutadiene will increase Tg, with a corresponding improvement in traction and wet grip performance. There would also be a corresponding drop in tread wear performance.

4. Addition of or increase in styrene concentration in SBR leads to an increase in Tg and improvement in traction performance. As an empirical guideline, there is a near-linear relationship between styrene and 1,2-vinyl butadiene levels in the polymer, with two 1,2-vinyl butadiene units providing traction performance equivalent to that of one styrene unit. This, in turn, allows

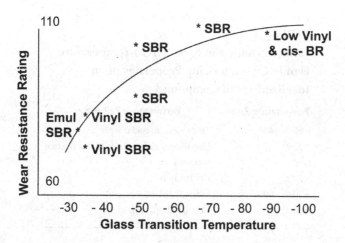

FIGURE 2.5 Tread Compound Tg and Tire Tread Wear (4)

optimization of the polymer microstructure to achieve a specific rolling resistance, traction, and wear performance, depending on the design and mission profile of the tire

5. The tread is the largest contributor to tire rolling resistance. The lower the compound Tg, the lower (better) will be the tire rolling resistance (Figure 2.6)

Such empirical observations allow the tangent delta temperature curve of a tread compound measured from −100°C to +100°C to be segmented into zones that would characterize the tread compound's performance (Table 2.1). Such property target definitions enabled the development of the concept of "integral rubber" or dual Tg

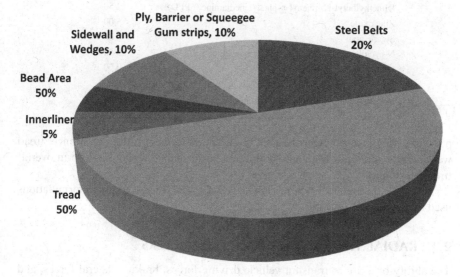

FIGURE 2.6 Tread is the Largest Contributor to Tire Rolling Resistance (4, 6, 7)

TABLE 2.1

Tangent Delta (tan δ, or G″/G′) Temperature Profile Characterizing Properties of an Idealized Tread Compound

Temperature Zone	Performance Parameter
−60 to −40°C	Tread wear, irregular wear
−20°C	Low temperature properties, ice traction
0°C	Wet traction
+20°C	Dry traction
+30 to +60°C	Rolling resistance
+80°C	Heat build-up, high speed

TABLE 2.2

Model Baseline Passenger Tire Tread Formulation (1, 5)

	PHR (parts per hundred rubber)
Emulsion SBR (1712 or equivalent with TDAE)	96.25
Polybutadiene	30.00
Carbon black (134)	80.00
Peptizer (if required)	0.25
Paraffin wax	1.50
Microcrystalline wax	1.50
Process oil (treated distilled aromatic extract, TDAE)	15.00
Polymerized trimethyl-1,2-dihydroquinoline (TMQ)	1.50
Dimethylbutyl-N′-phenyl-p-phenylenediamine (6PPD)	2.50
Zinc oxide	5.00
Stearic acid	2.00
N-cyclohexyl-2-benzothiazolesulfenamide (CBS)	3.50
Diphenylquanidine (DPG)	0.50
Sulfur	1.00
Retarder or pre-vulcanization inhibitor (PVI)	0.25

polymers; that is, a polymer that can be designed to meet rolling resistance, tread wear targets, and both wet and dry traction performance without a drop in overall tire performance.

Tables 2.2 through to 2.6 show some generic model tread compound formulations used in a variety of tires (5).

2.3   RADIAL TIRE TREAD DESIGN PARAMETERS

The ability of a tire to transmit vehicle driving forces, braking, lateral forces, and stability is influenced to a very large degree by the tire tread pattern (5). The design

TABLE 2.3
Model Performance Tire Silica Tread Formulation (1, 5)

	PHR (parts per hundred rubber)
Solution SBR	96.25
(25% styrene, 73% vinyl 1,2 butadiene, oil extended 37.5%)	
Polybutadiene (BR)	30.00
Silica (HDS)	80.00
Silane coupling agent (on carrier)	12.80
Paraffin wax	1.00
Microcrystalline wax	1.00
Process oil (treated distilled aromatic extract, TDAE)	10.00
Polymerized trimethyl-1,2-dihydroquinoline (TMQ)	1.50
Dimethylbutyl-N'-phenyl-p-phenylenediamine (6PPD)	2.50
Zinc oxide	4.00
Stearic acid	2.00
N-cyclohexyl-2-benzothiazolesulfenamide (CBS)	1.50
Diphenylquanidine (DPG)	2.00
Sulfur	1.50
Retarder or pre-vulcanization inhibitor (PVI)	0.25

TABLE 2.4
Model General Purpose Tire Tread Compound (1, 5)

	PHR (parts per hundred rubber)
Emulsion SBR (1502)	100.00
Carbon black (110, N121, N234)	50.00
Peptizer (if required)	0.25
Paraffin wax	1.00
Microcrystalline wax	1.00
Process oil (treated distilled aromatic extract, TDAE)	10.00
Polymerized trimethyl-1,2-dihydroquinoline (TMQ)	1.00
Dimethylbutyl-N'-phenyl-p-phenylenediamine (6PPD)	1.50
Zinc oxide	3.00
Stearic acid	1.00
N-cyclohexyl-2-benzothiazolesulfenamide (CBS)	0.85
Diphenylquanidine (DPG)	0.40
Sulfur	1.40
Retarder or pre-vulcanization inhibitor (PVI)	0.25

TABLE 2.5
Model Global Truck Tire Tread Compound (1, 5)

	PHR (parts per hundred rubber)
Natural rubber	100.00
Carbon black (N110, N115, N220, N234)	50.00
Peptizer	0.25
Paraffin wax	1.50
Microcrystalline wax	1.50
Process oil (treated distilled aromatic extract, TDAE)	3.50
Polymerized trimethyl-1,2-dihydroquinoline (TMQ)	1.50
Dimethylbutyl-N′-phenyl-p-phenylenediamine (6PPD)	2.50
Zinc oxide	4.00
Stearic acid	2.00
t-Butyl-2-benzothiazolesulfenamide (TBBS)	1.95
Diphenylquanidine (DPC)	0.35
Sulfur	1.10
Retarder or pre-vulcanization inhibitor (PVI)	0.25

TABLE 2.6
Model Off-Road Tire Tread Compound (1, 6)

	PHR (parts per hundred rubber)
Natural rubber	100.00
Carbon black (N220)	45.00
Silica	15.00
Peptizer	0.25
Natural resin	5.00
Paraffin wax	1.50
Microcrystalline wax	1.50
Process oil (treated distilled aromatic extract, TDAE)	3.50
Polymerized trimethyl-1,2-dihydroquinoline (TMQ)	1.50
Dimethylbutyl-N′-phenyl-p-phenylenediamine (6PPD)	2.50
Zinc oxide	4.00
Stearic acid	2.00
t-Butyl-2-benzothiazolesulfenamide (TBBS)	0.50
t-Butyl-2-benzothiazolesulfenimide (TBSI)	0.50
Sulfur	1.10
Retarder or pre-vulcanization inhibitor (PVI)	0.25

of a tread pattern consists of grooves molded from a smooth tread surface, thereby creating ribs which can then be further sub-divided into blocks depending on the tire application or vehicle wheel position for which the tire has been designed. This enables attainment of tire road grip, the dissipation of water to minimize hydroplaning and improves tire–vehicle steering qualities. The grooves in the rubber are therefore designed for expulsion of water from the tire-pavement footprint and thus to prevent hydroplaning. The ratio of the tread net contact area with the pavement to the total footprint area, sometimes referred to as the net-to-gross, is thus an important property in assessing tire performance.

The tire tread design also has an effect on noise generated, especially at speeds over 50 kph. For example, at high speeds, up to 80% of the noise generated by a vehicle can be attributed to the tires. The engineer must design tread patterns to not only meet the mission profile for the tire but also be aware of the service environment in which the tire will function, i.e., use on wet surfaces, low noise generation, high-speed capability, and all-season applications. Tires with a smooth tread (i.e., having no tread pattern) are known as slicks and are generally only used for high-speed racing; the lack of wet traction performance prevents their use on conventional road surfaces.

Tire tread design variables include:

1. Ribs
2. Grooves
3. Void area
4. Blocks or lugs
5. Shoulder slots
6. Block-lug element orientation
7. Number of notches
8. Tie bars
9. Sipes

A tire's tread ribs, when divided into blocks, can improve traction qualities. The tread elements are thus arranged in a series of repetitive patterns of voids, ribs, blocks or lugs, slots or sipes, and grooves. The tread pattern can be further defined in terms of length, width, percentage of void area within the tread pattern, lug or block dimensions, aspect ratio (defined as the block height divided by its length), and block wall angles. The pattern of blocks and lugs becomes important in high-torque applications, such as the drive axle position in heavy-duty trucks, or in high-performance passenger tires to achieve the necessary traction without loss in other properties. If the tread block aspect ratio is too high, the design may be more susceptible to abnormal or irregular wear patterns (Figure 2.7). Though such tread patterns show excellent traction performance, irregular block wear patterns can occur due to distortion and snap-back as the tread block rotates through the footprint at high frequency. Tread compound techniques to overcome trailing edge lug wear, sometimes referred to as "heel-and-toe" wear, involve increasing the compound dynamic stiffness or

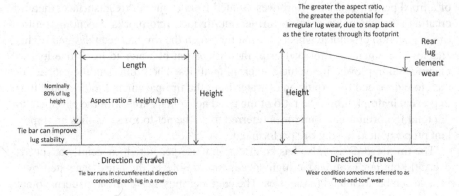

FIGURE 2.7   Tire Tread Lug Aspect Ratio (8)

storage modulus, G′. Some resins have been reported to impart such properties by increasing the compound Tg and storage modulus, and decreasing the Payne effect, thereby having little effect on hysteresis (8).

A similar situation is found in tires with an outboard row of small lugs (Figure 2.8, Pattern B). Such tires show excellent traction, though they can be susceptible to alternate step wear, sometimes referred to as "high-low lug wear". Such conditions can be addressed by adding resins or modifying the molds to add a tie-bar, which is a shallow connecting rib at the base of the lugs.

Tire tread elements are arranged to give optimum depth, tread width, and net-to-gross contact area. Many of the tread design principles are best described by

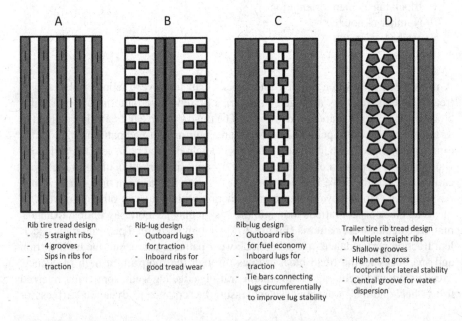

FIGURE 2.8   Illustrated Design Principles for Truck Tire Tread Patterns

considering truck and bus tire treads. For truck tires, tread patterns can be classified into five basic categories (9):

1. Highway straight rib, sometimes with sipes along the rib length for traction
2. Highway rib-lug configurations, where the rib can be either inboard or outboard in the shoulder area; the lugs will correspondingly be opposite the ribs, outboard or inboard, respectively. In long-haul operations, the trend has been to place the ribs outboard, i.e., solid shoulders, for improved fuel economy
3. Highway lug designs
4. On-/off-highway rib or lug
5. Off-highway all-lug design.

The varying requirements for tread wear, traction, and rolling resistance provide simple bases for tread design principles that are best illustrated in Figure 2.8, which shows the rib and lug design options available.

Figure 2.9 illustrates a selection of truck tire tread patterns described here (2). Rib designs, with the ribs principally in the circumferential direction, are the most common type of highway truck tire tread pattern and show overall good service for a broad range of conditions. Such rib patterns in the tread pattern design also show low rolling resistance and very good fuel economy. On commercial trucks, they are used almost exclusively on the steer axle and trailer axle positions because of their lateral traction and uniform wearing characteristics. City and intercity buses use similar tires. Rib-lug combinations tend to find use on all-season tires, which require a balance of wear, traction, and wet skid resistance. On heavy-duty truck drive axles,

FIGURE 2.9   Truck Tire Tread Patterns (2)

where forward traction is a prime requirement, and where fast wear can occur as a result of torque-induced slip, a highway lug design is required. For off-highway service conditions, the tread pattern assumes a staggered joint lateral circumferential direction for both lateral and forward directions. Grooves tend to be larger and deeper, with the rib walls angled to eject stones or other objects, preventing them from becoming embedded in the tread.

Steer axle and trailer axle truck and bus tires frequently have a modification in the shoulder, sometimes referred to as a decoupling or protection groove (Figure 2.10). The purpose is to prevent or arrest fast wear, irregular wear, or other abnormal wear patterns, which, in many instances, start in the shoulder. Such wear conditions can start at the shoulder because of lack of belt support under this part of the tread. The groove can have one of three possible configurations: i) approximately 5 mm inboard from the shoulder edge, ii) at the shoulder edge, or iii) recessed below the shoulder edge. The groove depth can result in 50% to 100% of the primary tire tread grooves depending on the tire tread compound tear strength. Low tear strength can result in cutting, tearing, and chunking of the groove resulting in a loss in performance and potential warranty claims. In the case of a sidewall-over-tread construction, where the sidewall extends up to the tread, then a recessed decoupling groove may help mitigate any tear damage due to the lower tear strength of the sidewall compound.

The base of the major grooves in these tires often have a row of small lugs, about 2 mm in height, called stone ejectors or a stone ejector rib (Figure 2.10). These micro-lugs, sometimes referred to as buttons, are very effective at preventing stones or other small items of road debris from becoming lodged in the tread grooves and damaging the tire tread crown area.

Similarly, like heavy-duty truck tires, automobile tire tread designs can be placed in some broad categories such as high-mileage tire tread patterns, all-season, or broad-market tires, traction, high-performance, and traction-high-performance.

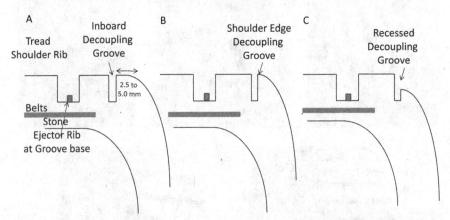

FIGURE 2.10 Truck Tire Tread Decoupling Groove Configurations. 1. Inboard and shoulder edge grooves (A, B) for duplex tread and base configurations, due to high tear strength of a typical natural rubber-based tread compound. 2. Recessed groove (C) for triplex tread base sidewall extrusions (Figure 2.1), due to lower tear strength of natural rubber polybutadiene sidewall compound blend

There are some simple guidelines that the design engineer would be aware of, such as the need for structural rigidity and aspect ratio of the tread blocks to prevent irregular wear, the use of tie-bars, and curvature at the base of tread elements to prevent high-frequency cyclic strains and consequent fatigue cracking. Given such general guidelines, tread pattern designs are still very diverse, with few published insights into pattern optimization. Table 1.3 has shown one possible scheme on how tire tread design is defined for a given tire performance level or mission profile (2). The Tire and Rim Association has prepared guidelines for tire design tread depth as the radial distance measured along the centerline of the mold, beginning at the outer cross section of the tread to the upper or outer section of the under-tread (10). In practice, the nominal tread depth is best reported as the depth from the top of the rib or lug at the centerline to the base of the groove nearest the centerline.

The trend in highway drive axle truck tire tread design is to have solid shoulders (Figure 2.9, Design 2) rather than outboard lugs (Figure 2.9, Design 4). Even though outboard lugs display better traction characteristics, such tread designs require compounds consisting of natural rubber, synthetic rubber, such as polybutadiene, and, as a consequence, high levels of carbon black to achieve the required tensile strength and, more importantly, fatigue resistance. Such compounds tend to display higher hysteresis (higher tan δ), resulting in higher rolling resistance and increased vehicle fuel consumption. In the case of tread patterns with closed shoulders (e.g., Figure 2.9, Design 2), all natural rubber compounds can be used with much lower hysteresis, and consequently better tire rolling resistance. Traction is then optimized by adjusting the dimensions of the central rows of lugs and the net footprint contact area.

The ratio of net road or pavement contact area to gross tread surface area ("net-to-gross") changes as wet traction becomes more important. Table 2.7 illustrates the net-to-gross percentage for typical commercial truck tire tread patterns illustrated in Figure 2.9. The table also lists some typical tread depths, defined as the distance from the top of the centerline rib to the base of the groove.

In determining the section height of tires, where the tread depths vary as in Table 2.7, the section heights are all equal to that of the steer axle tire plus the difference in tread height, rounded to the nearest millimeter. The objective is to minimize variation in new tire outer diameter by axle position. Including these design variables, the terms used to define tread patterns are listed as follows:

TABLE 2.7
Tire Tread Pavement Net Contact Area (1, 9) (Net-to-Gross)

Type of Service	Tread Pattern	% Net-to-Gross (Contact Area)	Nominal Tread Depth (mm)
Highway steer axle	Rib	72–78	15.0
Highway drive axle	Rib-lug	70–75	24.0
Highway trailer axle	Rib	75–85	10.0
On-/off-road	Rib-lug	65–75	20.0
Off-road	Lug	55–65	25.0

1. Groove amplitude: In staggered groove designs, this is the distance the groove pattern oscillates about the central or circumferential direction. It is analogous to a sine wave. Though an increase in amplitude improves traction, the trend is to minimize it, so as to reduce irregular wear patterns, rib edge wear, and depressed rib wear
2. Sipes: Small individual tread voids in the tread rib or block, typically added to improve traction performance
3. Pitch length: Length of each repeating unit in a tread pattern. Variable pitch lengths in a tread design are used to minimize noise generation. Pitch would be analogous to the wavelength of a sine wave
4. Blade: A protrusion on the tread mold that forms part of the tread design. The protrusion forms a corresponding depression in the finished tire
5. Stone ejection ribs: Position of the tread designed to prevent stones becoming trapped in the tread grooves. They appear as a row of minor lugs from 2–3 mm in height at the base of each groove
6. Tie bars: Ribs running circumferentially around the tire, connecting the base of a row of lugs, and helping to prevent the base of the lug cracking due to lug flexing and fatigue. Tie bar height can range from 25% to up to 60% of the lug height
7. Decoupling groove: A groove in the shoulder of the tire designed to prevent irregular wear and fast shoulder wear.
8. Tread wear indicator: Tie bar approximately 2–4 mm in height, running laterally across the tread to serve as an indicator when the tire will need to be removed.

As described in the Federal Motor Vehicle Safety Standard FMVSS 139, the tread wear indicators on a worn tire will appear as solid bands across the tread surface when the tread has a remaining depth of 1.6 mm ± 0.38 mm (0.0625 inch ± 0.0156 inch).

In conclusion, the Tire and Rim Association has developed extensive tables on tread depth guidelines (10). Table 2.8 shows a brief summary of some of the design

TABLE 2.8

The Tire & Rim Association Design Guidelines for Tire Tread Depth (10)

Tire Size	Highway Rib	Highway Deep Lug Traction
	[inches]	[inches]
6.00R16 LT	0.40	0.52
9.00R16 LT	0.49	0.61
11R22.5	0.60	0.81
11R24.5	0.60	0.81
12R22.5	0.62	0.84
10.00R20	0.51	0.69
11.00R22	0.53	0.71
445/65R22.5	0.728	0.965

tread depths. In practice, tires will have deeper tread depths in order to achieve the wear performance targets and traction requirements. However, it does provide a valuable starting point when designing a new tire.

Tires for passenger vehicles and light trucks will fundamentally fall into one of five broad design categories (Table 1.3);

1. Central solid rib with outboard rows of blocks
2. Block-ribs with the circumferential rows of blocks connected with a tie bar
3. All-lug or -block design
4. Central groove with outboard blocks
5. Directional, both symmetrical and asymmetric

Figure 2.11 illustrates examples of the five fundamental design patterns, each optimized for the vehicle classes on which they will be placed, or the market for which the tire is designed. In addition to tread pattern design, other elements that are important in tuning a tire for a vehicle fitment will include the tread compound, belt design (belt angle, wire and belt stiffness, gauges, and widths), and bead design (ply ending height and apex or bead filler properties).

Snow and winter tires are designed for use in colder weather, snow, and ice. Many designs adopt a directional tread pattern but have a different tread compound to facilitate grip at low temperatures. Though many properties are very important, the governing factor in snow or winter tire tread compound type will be the compound Tg (Table 2.9).

For other tire lines, the tread designs tend to be less complex. For example, in agriculture or farm equipment tires, the tractor front or steer axle tires are typically a simple rib design The rear drive tires are the classic directional lug design, whereas implement tire treads have rib or sometimes rib-button configurations. Aircraft tires,

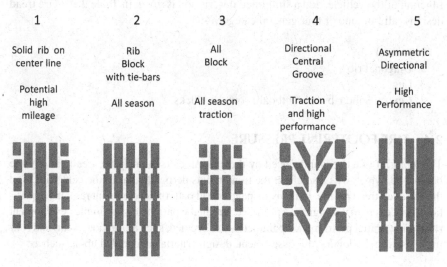

1	2	3	4	5
Solid rib on center line	Rib Block with tie-bars	All Block	Directional Central Groove	Asymmetric Directional
Potential high mileage	All season	All season traction	Traction and high performance	High Performance

FIGURE 2.11   Illustrative Examples of Passenger Tire Tread Patterns: Characterization and Potential Applications

TABLE 2.9

Passenger Tire Tread Compound Nominal (starting point) Tg Target by Application

Tire Application	Compound Tg Target	Typical Polymers
Snow, winter	−50°C	NR/BR Blend
All-season	−30°C	SBR/BR
Base-line	−30°C	SBR/BR
Performance	−20°C	High styrene/vinyl BD Blends
High Performance	−10°C	High styrene/vinyl BD Blends

TABLE 2.10

Earthmover Tire Generic Tread Patterns

Application	Front Steering Axle	Rear Drive Axle
Backhoe	Rib	Direction lug (similar to farm)
Loader	Lug	Lug, diagonal rib
Truck	Lug	Lug, rib-lug
Grader (highway construction)	Lug	Lug, rib-lug
Scraper (pavement construction)	Lug	Lug, rib-lug
Dozer	Tracks	Tracks

with few exceptions, are all of a rib design for rapid water dispersion to prevent skidding upon landing. Earthmover tires fall into one of six categories defined by the function of the vehicle, and a simplified description is given in Table 2.10. Tire tread designs fall into one of four general categories (1),

1. Lateral rib
2. Diagonal rib
3. Lug
4. Central solid rib with outboard rows of blocks.

2.4   TIRE FOOTPRINT PRESSURE

The footprint of a tire is influenced by three parameters, namely the tread pattern, the belt lay-up, which will also affect the tread radius perpendicular to the centerline, and the extruded tread profile. In many respects, tread pattern development remains a qualitative process in that designs are proposed within the parameters of simple development rules. First, initial performance parameters are decided, such as traction, wear, and rolling resistance. Following this assessment, design criteria are decided upon, such as:

- Net contact area
- Groove depth (within industry guidelines)

- Lug aspect ratios (too high a ratio can lead to a "heel-and-toe" or "high-low" wear phenomenon)
- Rib amplitude and width variations. High circumferential amplitude of the rib around the tire could lead to accelerated depressed rib wear or irregular wear in sections of the rib, creating a constraint in the design process.

Inflation pressure also has a large effect. Low pressure can accelerate wear rate, in addition to inducing a loss in tire durability, due to increasing pressure in the shoulder region, but less pressure in the centerline, rendering it susceptible to scrubbing.

Upon identifying potential designs, experimental molds can be produced, and test tires produced for evaluation. Alternatively, the tread pattern can be carved on a smooth tire tread and the tire then tested. The evaluation will then follow a testing protocol defined by industry standards such as QS9000 or TS16949. Each tire company has a unique testing protocol, so only a simplified or "high-level" guideline based on these industry standards is discussed later (Table 2.11 shows variables impacting traction).

The belt design and lay-up, which can also affect footprint pressure, will be discussed along with the other aspects of tire casing construction.

2.5   TREAD EXTRUSION CONTOUR

With regard to the extruded tread contour or profile, this is set at the initial tread extrusion operation. For illustrative purposes, a truck steer axle tire, which is typically a five-rib tread, has a tread extrusion profile which would follow the ribs and grooves in the tire mold (Figure 2.12). The depth of the channels and the height of the humps, corresponding to the ribs, will ensure the compound flows to fill the voids in the mold tread pattern. Inadequate compound flow can be seen in the final cured tire and are sometimes described as "lights". In more severe instances, inadequate compound flow can cause reduced footprint pressure in the affected rib, and, in turn,

TABLE 2.11

Factors Affecting Tire Traction Performance (11)

Variable	Elements
Tire	Inflation pressure
	Casing stiffness
	Tread pattern groove dimensions
	Groove depth
	Tread block dimensions
Pavement	Micro-texture
	Macro-texture
	Porosity
Fluid medium (surface oils)	Viscosity
	Density
	Depth
Vehicle	Speed

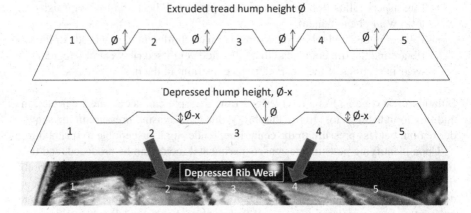

FIGURE 2.12 Extruded Tread Compound Profile Corresponding to a Five-Rib, Four-Groove Tread Pattern. Longitudinal view of the tread and depressed rib wear (12)

fast wear due to longitudinal scrubbing. Figures 2.12 and 2.13 illustrate a case where the hump height, corresponding to the 2nd and 4th ribs of an extruded tread profile, is too low. The resulting low footprint pressure can then result in scrubbing and, in turn, fast wear, leading to 2nd and 4th depressed rib wear. Though other factors, such as suspension defects, poor tire bead seating, low inflation pressure, and light loads, can induce this condition, in the event of warranty claims it is an item for the tire design engineer to check.

In cases where the tire shoulder footprint pressure is low, this could lead to scrubbing and fast shoulder wear. This can sometimes be predicted by evaluating new tire footprint pressure (Figure 2.13). Low shoulder pressure can be caused by several tire design parameters, such as low shoulder hump height, narrow shoulder ribs,

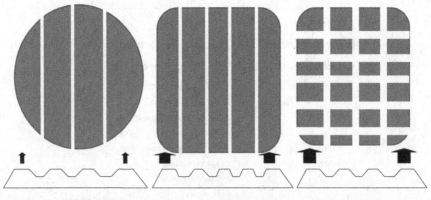

Low shoulder footprint pressure leads to fast shoulder wear   High shoulder footprint pressure leads to best shoulder wear performance   Another example of good footprint pressure

FIGURE 2.13   Tread Extrusion Illustration and Potential Impact on Tread Wear Performance

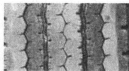

Diagonal wipe	Depressed 2nd & 4th rib wear	Fast shoulder wear (right side)

FIGURE 2.14 Samples of Tire Abnormal Wear Conditions (13) [Many times due to vehicle alignment settings]

non-linear shoulder ribs with high inboard rib amplitude (i.e., straight outboard rib wall but oscillating inboard rib wall), and narrow belts, providing inadequate support. However, as in the case of depressed rib wear, other factors can create this condition, such as vehicle alignment, condition of the suspension, bent axles, and light vehicle loads.

In summary, abnormal wear patterns could be classified as one of three types: irregular wear such as diagonal wipes and depressed lug wear, accelerated rib wear, and fast shoulder wear (Figure 2.14). Though tire design and manufacturing can account for some of these conditions, the vehicle condition also has a major impact on tire wear conditions and will be discussed below (12).

2.6 TREAD RADIUS

The final point in tread and tire crown area design would be the tread radius, this being determined by the mold shape, belt width, and extruded tread contour (14). Though some lateral curvature on the tread will assist in achieving vehicle steering precision, excess curvature, i.e., when r_2 is significantly lower than r_1 (Figure 2.15), fast shoulder wear will result.

For truck and bus tires, such as a 275/80R22.5 size, nominal tread radii are in the order of 60 cm or more to minimize such fast shoulder wear. When the tread radius,

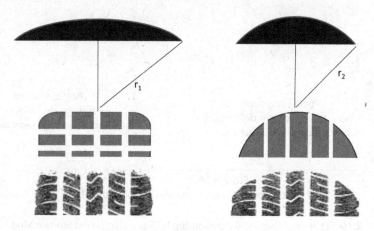

FIGURE 2.15 Tread Radius and Tire Footprint

measured in the lateral direction, drops toward 45 cm, in many instances fast shoulder wear could be observed.

2.7 TREAD WEAR MECHANISMS

The mechanism of tire tread wear is not fully understood. Given this, one fundamental interpretation is that wear occurs in one of two ways: i) thermo-oxidative degradation of the tread compound observed under slow-wearing conditions, or ii) a tensile strength-tearing phenomenon seen with faster wear rates. In the latter case, when soft rubber is abraded by repeated sliding or tearing, a series of ridges can be observed at right angles to the direction of travel. Such patterns have become known as Schallamach waves. For hard rubber compounds, indentations parallel to the direction of travel can occur, being created by erosion and friction.

Irregular wear patterns and environmental factors affecting wear and wear rates are essentially subsets of these two mechanisms. Factors affecting both tire global and irregular wear are:

1. Compound stiffness and storage moduli in both shear (G') and tension (E'), illustrated in the schematic Figure 2.16
2. Tread element dimensions such as tread lug aspect ratio
3. Groove and rib amplitude
4. Rib gauge variation
5. Belt design (width and stiffness)

The vehicle will also have an effect on tire wear. Fast wear and abnormal wear can be the result of vehicle mechanical problems or the unsuitability of the operating terrain. Maintenance and operational concerns would include:

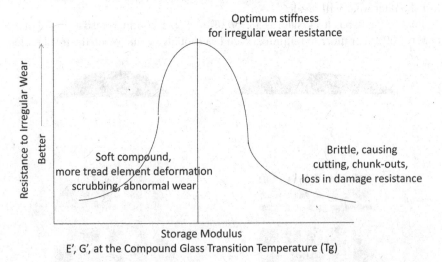

FIGURE 2.16 Proposed Empirical Relationship between Compound Storage Modulus and Irregular Wear (8, 16)

1. Improper tire mounting and bead seating
2. Adjacent axle alignments
3. Wheel misalignment
4. Mismatch of tires on dual drive and trailer axle positions, tire misapplication, and use of an inappropriate tire for the axle position and service conditions
5. Inappropriate or damaged suspension components, inadequate maintenance, broken or worn suspension parts, -wrong suspension compliance or spring rate, and incorrect replacement parts
6. Incorrect inflation pressure
7. Defective brakes creating flat spots in the tread
8. Overloading
9. Excessive imbalance and run-out, and non-uniformity of the rotating tire and wheel assembly
10. Driver practices

The road surface can impact wear rates. Depending on the age of the road, surface wear rates will change due to changes in road surface texture, profile, and grade or slope from the road centerline to the edge, and polishing effects. Temperature and weather also have an effect (14).

Tread wear is therefore affected by many complex variables. For example, weather and environmental changes can have an influence. In a study, seasonal wear data were collected for long-haul bus tires, which were mounted at various times throughout the year. Tires mounted in the autumn or fall operated through one winter, one summer, and then another winter. Other groups of tires were mounted in summer months and ran through only one winter period. Those tires which ran through two cold seasons showed slower wear rates *versus* those that underwent two hot summer service periods (Figure 2.17). From such studies

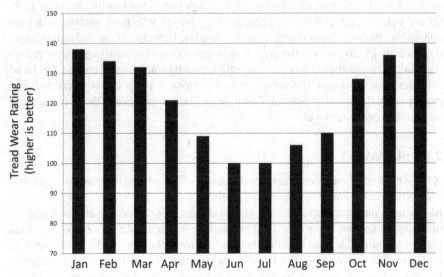

FIGURE 2.17 Seasonal Effects on Tire Tread Wear Rate (15)

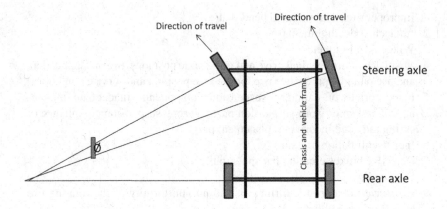

FIGURE 2.18 Ackerman Angle (17)

on season and temperature, temperature coefficients were proposed with natural rubber-based compounds being more susceptible to faster wear rates, compared with emulsion SBR. From linear data relationships, temperature coefficients were calculated and, for natural rubber, were in the order of 3% per °C for natural rubber and 1.5% per °C for SBR. Depending on the test conditions and operating environments, there is the potential for reversal in wear rankings, thus creating additional complexity in tread wear studies (9, 15).

A second example is the impact of the Ackerman angle (17) or the relative angle between the left and right steering axle wheels during a turn. In practice, this compensates for the difference in turning radius between the two front wheels, thereby minimizing lateral tire scrubbing and wear (Figure 2.18). The Ackerman principle is complex and further reference is recommended, particularly in solving steer axle irregular wear concerns, should they arise. A low turning radius for inner city pickup and delivery vehicles is much preferred by drivers who must maneuver in tight conditions, thus necessitating tight turning angles. However, if the Ackerman angle (Ø) exceeds 15–20°, excess tire wear can be generated, leading to significant reductions in tire removal mileage (7). In addition, misalignment would cause a fast tread shoulder wear condition. If detected early, alignment settings can be corrected, and the tire rotated to another position which, in practice, could enable the tire to achieve its design miles to removal.

2.8 SUMMARY

A large amount of the research and development effort which tire companies put into improving performance is committed to tread technology. Though much research has focused on computational tools and modeling, in practice, for all tire lines, the major variables that impact the final product performance are the compound, tread pattern, tread profile, and tread radius, and, in turn, the tire–road contact patch or footprint. This necessarily brief review, which has included practical experience, has defined the key areas to be addressed in ensuring the tire can meet the mission profile for which it has been designed.

REFERENCES

1. Rodgers B, Waddell W. The Science of Rubber Compounding. In *Science and Technology of Rubber.* Eds. JE Mark, B Erman, CM Roland. Academic Press, Boston, MA. 2015, 653–696.
2. Rodgers B, D'Cruz B. Tire Engineering. In *Rubber Compounding Chemistry and Applications.* Ed. B Rodgers. CRC Press, Bota Raton, FL. 2015, 279–599.
3. Rodgers B, Waddell W, Klingensmith W. *Rubber Compounding. Kirk-Othmer Encyclopedia of Chemical Technology,* 5th Edition. John Wiley & Sons Inc., New York. 2003.
4. Rodgers B, Tracey D, Waddell W. *Tire Applications of Elastomers, 1. Treads.* Paper H, Presented at a Meeting of the American Chemical Society Rubber Division, Grand Rapids, MI. 2004.
5. Rodgers MB *Tire Formulary.* ELL Technologies LLC, Cedar Park, TX. 2019.
6. Waddell W, Rodgers B. *Tire Applications of Elastomers, 2. Casing.* Paper I, Presented at a Meeting of the American Chemical Society Rubber Division, Grand Rapids, MI. 2004.
7. Waddell W, Rodgers B. *Tire Applications of Elastomers, 3. Innerliners.* Paper J, Presented at a Meeting of the American Chemical Society Rubber Division, Grand Rapids, MI. 2004.
8. IP.com Prior Art Database Technical Disclosure, Anonymously. *Compounding Method to Improve Irregular Wear Resistance of Tire Tread.* IP.com Number IPCOM000249914D. April 28, 2017.
9. Ford T, Charles F. *Heavy Duty Truck Tire Engineering.* Society of Automotive Engineers, SP-729, Warrendale, PA. 1988.
10. *Engineering Design Information.* The Tire and Rim Association Inc., Copley, OH. 2005.
11. Rodgers B, Mezynski S. *Radial Medium Truck Tire Performance and Materials.* Presented at a meeting of the American Chemical Society, Rubber Division, Akron Rubber Group. 1988.
12. Technology and Maintenance Council. *Radial Tire Conditions Analysis Guide,* 3rd Edition. 2004.
13. Lindenmuth BE. An Overview of Tire Technology. In *The Pneumatic Tire.* Eds. AN Gent, JD Walter. U.S. Department of Transportation, National Highway Safety Administration, Washington, DC. 2005.
14. Kovac F. *Tire Technology,* 5th Edition. The Goodyear Tire & Rubber Company, Akron, OH. 1978.
15. Schallanmach A, Grosch K. Tire Traction and Wear. In *Mechanics of Pneumatic Tires,* Ed. SK Clark. U.S. Department of Transportation, National Highway Safety Administration, Washington, DC. 1975.
16. Rodgers MB ELL Technologies LLC Data. Private communication. 2020.
17. Fitch JW. *Motor Truck Engineering Handbook.* SAE, Warrendale, PA. 1994.

3 Tire Casing Construction

3.1 INTRODUCTION

The construction of a tire is best viewed as two parts, namely the tread or crown area, consisting of the tread components, which has been considered in Chapter 2, and the casing or carcass, which consists of the belts and associated components, such as the overlay, compound wedges between the belt edges, the ply, sidewalls, bead area, and the inner liner (1). With respect to the casing, two enabling technologies have allowed the development of the performance envelope observed in modern radial tires: i) the steel wire belt design, facilitating a flat footprint and, in turn, improved tread wear resistance with high removal mileage and good traction, and ii) the inner liner, made of halobutyl rubber, which has allowed improvements in durability and fuel economy by maintaining consistent inflation pressure. The quality of the inner liner has had a major impact on the durability of a tire and is covered in more depth in Chapter 4. The tire casing can therefore be considered as four elements, namely the belt lay-up, ply and sidewall, the bead construction, and the materials and compound line-up.

3.2 CASING COMPONENTS

Bias and radial tires have many components in common, and, in some instances, the components use similar compound formulations. The casing, illustrated in Figure 3.1, consists of the following parts, listed from the crown of the tire to the bead.

Overlay: Fabric layer between the belts and the tread, intended to constrain the centrifugal forces when the tire is rotating at high speed. It may be full width, spiral-wound fabric-reinforced strips, or belt edge cover strips

Belts: Typically produced from brass-coated steel wire. Passenger tires (PCR) have two belts, and truck and bus radials (TBR) would typically have four belts, whereas larger off-road earthmover tires ("off-the-road", OTR) can have up to six belts

Belt Wedges: Components placed between adjacent belts and at the belt edges to help maintain a flat footprint when the tire is under load. An example is the 2/3 wedge in a truck tire, which is the wedge at the end of Belts #2 and #3, stabilizing the belt ending.

Shoulder Wedge: Triangular-shaped component located at the edge of the Belt #1 and between the ply and the sidewall

Sidewall: Component extending from the bead area to the tire shoulder, to improve tire aging resistance and to protect the tire from curb scuffing

White Sidewall: Ranges from a full white-colored sidewall, bead to shoulder, a white stripe, or only white lettering. All would have the same compound

FIGURE 3.1   Truck Tire Casing Components (2)

formulation based on chlorobutyl rubber, Ethylene Propylene Diene
Monomer (EPDM) synthetic rubber, and natural rubber

Sidewall Insert: Compounded rubber or a calendared fabric strip extending
from the bead region up the sidewall

Sidewall Scuff Rib: Layer of rubber in the sidewall to protect it from severe
curb scuffing. It may also have wear indicators to show the degree of sidewall abrasion

Sealant: Component designed to block an air leakage in the event of a puncture

Ply: Calendared fabric or wire which reinforces the carcass or casing. The calendared compound is referred to as wire coat, ply coat, or skim compound

Inner Liner: A critical strip of rubber on the inside of the tire, extending from
one bead to the opposite side bead, and ensuring that the tire maintains its
inflation pressure

Barrier: Also termed a squeegee or tie gum, it provides an adhesive layer and
reinforcement layer for the inner liner

Bead: A steel hoop made from a brass or bronze spiraled wire. It is coated with
a compound termed the bead insulation during the bead building operation

Bead Wrap: Fabric wrap to hold the bead cables together during tire manufacturing

Apex: The triangular component above the bead and extending 25% to 30%
up the sidewall. It is also referred to as the bead filler and, depending on the
tire size, can be just one compound (in passenger tires) or two compounds,
termed Apex 1 and Apex 2, as in truck tires. The function is to provide a
modulus gradient up the sidewall, thereby preventing concentration of a
stress and creation of a fracture point

Chafer: A hard rubber compound which helps lock the tire onto the rim and
prevent rim-tire abrasion. It can also be referred to as a toe guard (in passenger tires)

Flipper: A fabric strip located between the bead and the ply, preventing the
two components from damage due to abrasion as the tire rotates under load

Gum Strips: These are different types of rubber strips, all made of the same compound, located at belt endings, ply endings, and other component endings, and serving as a transition compound between two different tire parts. They provide a modulus gradient, improve component-to-component adhesion, and minimize potential for crack formation and propagation

The formulations of many of the casing compounds differ, adding to the composite complexity, but, as a general guideline, components will have some common characteristics to improve manufacturing efficiency. Table 3.1 shows the major polymers and carbon blacks used in various compounds. Many tire companies have consolidated carbon blacks to five or six grades on the grounds of efficiency. This is important as it reduces the number of towers and carbon black-handling systems required at the manufacturing plant. Formulations can then be tuned according to intended use, using polymers, process oils, and the vulcanization process.

There are a number of fundamental functions which the tire casing must be able to carry out to meet structural requirements, which are:

1. Resistance to significant change in inflated tire dimensions with increasing inflation pressure or over time at a given inflation pressure;
2. Ability to envelop or roll over road obstructions without sustaining damage;
3. Ability to deform from a surface of double curvature to a flat plane without damage; and
4. Ability to be sufficiently rigid that it can develop forces in any direction, as under steering conditions.

TABLE 3.1

Polymer and Carbon Black Types by Casing Component

Component Compound	Natural Rubber	Synthetic Rubber	Typical Carbon Black
Apex (bead filler)	TSR20		N550
Barrier	TSR20		N326, N347
Bead insulation		E-SBR	N330
Belt wedges	TSR20		N326, N347
Chafer	TSR20		N330
Fabric ply compound	RSS2, TSR10	E-SBR	N325, N347
Gum stripes	TSR20		N325, N347
Inner liner		BIIR, CIIR	N326
Sealant		IIR	N660
Shoulder wedge	TSR20		N550
Sidewall	RSS2, TSR10	—	N330
White sidewall	TSR10	CIIR, EPDM	—[1]
Steel ply compound	RSS2, TSR10	—	N325, N347[2]
White sidewall	TSR10	EPDM, CIIR	—

[1] White sidewall fillers consist of titanium dioxide, silica, and calcium carbonate
[2] Ply compounds may also contain silica to improve adhesion and reduce hysteresis

Such performance parameters account for the structural complexity of the modern radial pneumatic tire.

3.3   BELT DESIGN

The belts in a tire are located just below the tread components and consist of two or more layers of calendared wire laid down at opposite angles to each other and at a bias angle to the crown center line of the tire. Four factors determine the performance of a tire belt lay-up:

1. Belt wire angle;
2. Wire gauge and construction;
3. Belt composite gauge;
4. Compound stiffness.

Passenger tire belt angles, i.e., the direction from the center line, tend to be between 21° and 24°. Motorcycle tires differ from other tires in that, in this instance, belts are more circumferential, to produce a rounded footprint for good maneuverability, stability, ride, and handling.

As a general guide, for equal belt widths, the greater the belt stiffness, the better the potential tire wear performance achieved, by providing a solid foundation for the tread compound. Tire belt stiffness is best illustrated when considering truck tire constructions. Belt rigidity can be described by the term, "Gough Stiffness", which is a measure of the in-plane bending stiffness, or bending moment, of the belt wire, cord and rubber composite laminate (3, 4, 5). Dr Eric Gough, who was a researcher at The Goodyear Tire & Rubber Company, determined that a simple beam model of the tire belt construction could be used to evaluate relative tread wear performance. Using both shear and bending moments in the analysis, Gough derived an empirical equation, defining the Gough stiffness parameter (S) for a simple laminate:

$$S = P / d \qquad (3.1)$$

where load, P, is the force applied to the laminate or composite belt system to create a deflection, d. For a tire, the Gough stiffness parameter would be a function of the circumferential modulus and shear modulus of the belt composite,

$$S = (E \times G) / (C_1 \times E) + (C_2 \times G) \qquad (3.2)$$

where S is the Gough stiffness, E is the circumferential modulus of the cord and rubber laminate, G is the shear modulus of the laminate, and C_1 and C_2 are constants related to the tire dimensions or size. Equation 3.2 can be rearranged to a simple model, consisting of a simple supported beam of length, L, stress and strain elastic constants E and G, which is deflected a distance of d by a force P:

$$S = PL^3 / 48L + 2PL / 8 \, AG \qquad (3.3)$$

where A is the in-plane cross-sectional area, to which the force P is applied, and I is the moment of inertia of the beam.

In practice, a heavy-duty truck tire, for use on long-haul trucks and buses, typically consists of four belts numbered 1 through 4. Belt #1, by convention, is the first one applied at the tire building machine and will be laid down in a right- or left-hand direction to the center line. Belt #4 will be the last one applied and, similarly, will be in a right- or left-hand direction relative to the center line. Belt #1 is called the transition belt, and the belt angle at which it is laid down is in between that of the ply, and 90° to the center line and Belt #2, so as to help distribute and alleviate stress on the shoulder. The first belt can extend across the width of the crown of the tire or, in some instances, it can be split in two, i.e., a *split #1*, with the central void being filled with a gum strip compound, similar in composition to the belt compound formulation. Belts #2 and 3 are termed the working belts and are laid down at angles opposite to each other, whereas Belt #4 is termed a "protector belt", which helps to minimize belt damage due to road hazards and punctures.

As the angles of Belts #2 and #3, laid down in equal but opposite directions to the center line of the tire building machine, change, so will the Gough Stiffness parameter. Figure 3.2 shows Gough Stiffness data calculated for a series of belt composites, where the angles of Belts #2 and #3 ranged from 12° to 24° and were then extrapolated through to 26°.

For a belt lay-up of a right/right/left/right configuration, the Gough Stiffness parameter passed through a maximum value when Belts #2 and #3 were at approximately 15°. Over the past few years, the trend has been to reduce the belt angles toward 15° from, at one point, being around 23°, primarily to improve wear but also to reduce tire rolling resistance (Tables 3.2 and 3.3). The lower wire belt angle increases the radial stiffness of the tire and, consequently, decreases the deformation at a given load and inflation. Because the circumferential stiffness is increased,

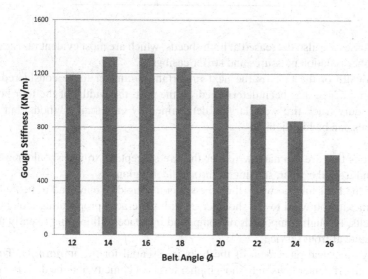

FIGURE 3.2   Gough Stiffness for a 4-Belt Truck Tire (5)

TABLE 3.2
Steer Axle Tires Measured Belt Angles (6)
(Average of 5 tires, size 275/80R22.5,
295/75R22.5. Tire dimensions were equal)

Belt #	Angle Off-Centerline	Angle Off-Centerline
	1980's to early 1990's	~2000's to current
1	65° right	50° right
2 ·	23° right	17° right
3	23° left	16° left
4	21° right	15° right

TABLE 3.3
Trailer Axle Tires Measured Belt Angles
(6) (Average of 5 tires, size 275/80R22.5,
2010 production onwards)

Truck Tire (TBR)

Belt	Angle Relative to the Centerline
1	50° right
2	22° right
3	16° left
4	16° right

inertia losses are also decreased at high speeds, which are most evident in truck tires with higher inflation pressures and stiffer casings.

The width of the belts is the next important parameter to be considered after optimum stiffness has been determined. In the past, the widths of the four belts in a heavy-duty truck tire were largely determined by empirical methods, but some guidelines can be highlighted;

1. Belts that are too narrow render the tire susceptible to fast shoulder wear and irregular wear, starting in the shoulder region
2. If the belts are too wide, then the structural rigidity may lead to belt edge cracking outward to the shoulder or with the cracks propagating along the belts, through compounds or compound interfaces, all leading to early tire removal from service
3. Symmetrical placement of the belts is essential for tire uniformity. Tires with off-center belts may show higher levels of conicity and the drivers may experience poor steering precision

TABLE 3.4
Truck Tire Belt Widths (6)
(Average of 6 tires, size
275/80R22.5, 295/75R22.5)

Belt	Width (cm)
1 Transition belt (full width)	18.5 +/- 0.1
2 First working belt	21.0 +/- 0.5
3 Second working belt	18.5 +/- 0.5
4 Top (protector) belt	8.6 +/- 0.1

4. Misplacement of the 2/3 wedge could cause flared belt endings and increasing belt edge strains
5. In the event of off-center belts and subsequent crack formation, the cracks would typically propagate between the two working belts
6. In the case of 3-belt constructions, with no transition belt, an increased belt angle, toward 25° or higher, of the first working belt would keep belt-ending stresses to a minimum.

Though every manufacturer is different, typical or nominal belt widths for a low-profile heavy-duty truck tire are tabulated in Table 3.4 for illustrative purposes.

Finite Element Analysis (FEA) has provided a quantitative method by which cyclic stress can be estimated and, in turn, by which the optimum belt lay-up can be calculated, so as to minimize the impact of cyclic strain on fatigue. The stress and strain at the belt edge of a truck tire can be calculated at three stages or conditions: i) inflated and unloaded, ii) when loaded at the center of the tire footprint, where greatest deformation occurs, and iii) when loaded but 180° away from the footprint center. Three points of interest might be the Belt #2 edge, the Belt #3 edge and the outside edge of the 2/3 wedge (Figure 3.3). From FEA, the belt edge stresses can be estimated (Table 3.5).

Though the stress at the belt endings might be low, considering a loaded vehicle travelling at 80 kph, the cyclic strains will approach 11 Hz, and fatigue or crack initiation conditions could develop, particularly in the event of belt misplacement, off-center belts, or waving non-uniform belt placement, sometimes referred to as "snaking". In cases where the inner liner allows oxygen permeation into the casing, thermo-oxidative degradation and fatigue may have an impact on the casing durability.

3.4   CASING CONSTRUCTION

Underneath the belts, the next important component is the ply, made of either fabric cords or wire, laid in a radial direction and extending from one bead to the opposite. The bead holds the ply in place in order that it can serve to reinforce the casing. In addition to locking the tire on the rim, it helps prevent rim-slip under high-torque conditions.

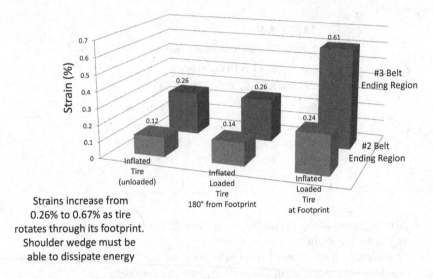

Strains increase from
0.26% to 0.67% as tire
rotates through its footprint.
Shoulder wedge must be
able to dissipate energy

FIGURE 3.3 Example of Belt Stresses and Strains in a Truck Tire as It Passes Through Its Loaded Footprint (4, 5)

TABLE 3.5
Strain Energy Density (MPa) Calculated by Finite Element Analysis

Location	Inflated Unloaded	Center of Footprint	180° From Footprint
	MPa	MPa	MPa
Belt #2 Ending	0.06	0.19	0.06
Belt #3 Ending	0.26	0.67	0.13
2/3 Wedge Ending	0.30	0.09	0.02

The tire design initially begins with the determination of the inflated dimensions of the required tire size. By use of inflated and growth characteristics of the tire, preliminary ply-line and mold dimensions are computed (Figure 3.4). The principles of ply-line determination were developed by John Purdy, a mathematician at The Goodyear Tire & Rubber Company, who derived the basic mathematical equations for cord path and tire properties (5). Purdy's equation, shown in Figure 3.4, defines the ideal inflated or "natural" shape for a "thin-film" structure. C=curvature of the ply line, Pc=radius from the center of the axle to the center of the ply-line or tire centerline, and Pm=radius from the center of the axle to the center ply- line width.

Once the mold boundary dimensions, location of the ply-line, and tread width and depth are known, the contours of the tread, shoulder, sidewall, and bead components can be determined. The primary interest in optimizing the design of a tire then lies in the belt area, bead area, and the belt and ply cord tensions. As the radial tire will contain multiple belts, these lay-ups must be viewed as a package (Figure 3.5).

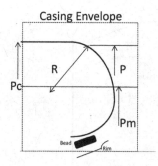

Casing Envelope

Ply line definition
Purdy's equation defines ideal inflated
natural shape for a thin film structure where;

C = Curvature of the ply line
Pc = Radius from the center of the axle to the center of the ply line or tire center line
Pm = Radius from the center of the axle to the center of the ply line width

Then:
$$C = 1/R = 2p / (Pc^2 - Pm^2)$$

FIGURE 3.4 Ply Line Definition (4, 5)

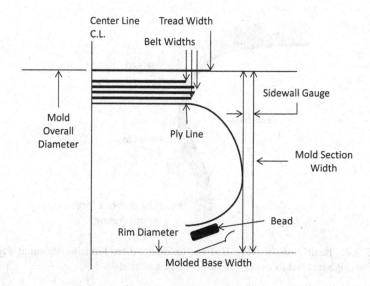

FIGURE 3.5 Ply Line Boundaries (4, 5)

If one considers an inflated non-belted tire, the ply cords, whether they are of a single-ply or multiple bias-ply construction, will assume a configuration which minimizes strain within the composite. The resulting cord path is termed the *neutral contour*. Due to the stiffness of the belt package, the belted tire has a restriction to the inflated diameter of the tire, and the neutral contour or ply-line of such systems is consequently altered.

As with the belt region of the tire, the ply reinforcement will also undergo varying stresses. Figure 3.6 illustrates the results of a computational analysis (Finite Element Analysis, FEA), showing the stress at the ply ending as the tire makes a full rotation and passes through the footprint.

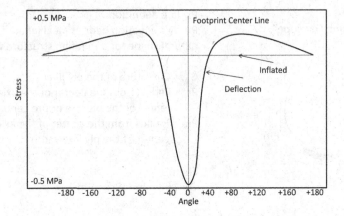

FIGURE 3.6   Impact of Ply End Stress as Tire Rotates through the Footprint (4)

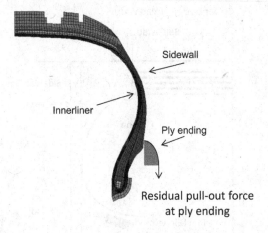

FIGURE 3.7   Bead Area and Ply Ending Residual Force, Inducing Potential Ply End Separation [high residual forces can cause ply end separations]

As the tire is inflated and loaded, there is a tendency at the ply ending for the reinforcement to bend outwards. This is particularly true for wire plies. This is termed a residual force and, if the ply ending is too low or the adhesion and tear strength of the compounds is not adequate, a ply-end separation and exterior crack can occur, eventually leading to the early removal of the tire from service (Figure 3.7).

Typical inflated cord tension plots for a truck tire are illustrated in Figure 3.8. In an unloaded state, the cord tension for the belts tends to be at the tire centerline, and the ply tension is greatest at the point corresponding to the sidewall location; however, on application of a load to the inflated tire and consequent deflection, the cord tensions increase at the belt edges away from the centerline and also in the bead zone. As reviewed earlier, these two regions tend to be the failure zones in a tire construction (Figures 3.8 and 3.9).

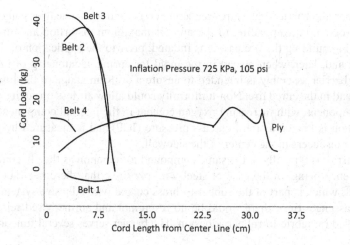

FIGURE 3.8 Inflated Unloaded Tire

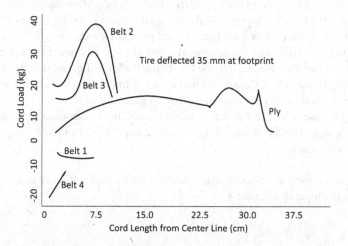

FIGURE 3.9 Inflated Tire Loaded and Deflected

3.5 INNER LINER AND BARRIER

The first component applied to the tire at the building machine is the inner liner, frequently with the barrier and squeegee (tie gum) already pre-assembled with the liner. Many tire companies use the terms barrier, squeegee, or tie gum interchangeably as all essentially mean the same component. In this text, the term "barrier" will be used for consistency. The liner should run from the toe of one side of the tire to the opposite side. In instances where the liner ending extends up the inside tire sidewall, sometimes due to building machine design limitation, air migration can occur through the lower bead area, which can penetrate up the ply cords to the shoulder region, reducing the tire durability due to thermo-oxidative or aerobic degradation. Tire inflation pressure retention (IPR) is a defining quality for many premium tire manufacturers. In nearly all instances, the lower the tire inflation pressure loss rate (IPLR), the better will be the performance of the tire.

The liner and barrier gauge are frequently contoured with a heavier gauge in the shoulder region, corresponding to the area of maximum distortion and stretching when during building the tire casing is inflated, prior to the application of the belts and tread and, later, when placed in the mold for curing. Contouring of the inner liner and barrier assembly is intended to ensure a uniform gauge of the liner from bead to bead in the cured tire. Non-uniformity could allow air loss through a portion of the component, with increasing oxygen content in the internal components. This phenomenon is known as intra-carcass pressure (ICP) and is measured by placing pressure transducers in the center of the sidewall.

The barrier is typically of the same compound formulation as the ply compound, the difference being, in the case of steel wire ply tires, that there may not be any cobalt salt, which is part of the rubber-to-brass coated wire adhesive system. In the case of passenger tires, there would be no resorcinol and amine-based reinforcing resin needed for fabric-to-rubber adhesion. The barrier serves several functions:

1. Adhesive layer between the halobutyl inner liner and ply;
2. Improved resistance to ply cord strike-through, i.e., when the tire is shaped during the building process, if the ply cord tension is too high, cords could pull through the inner liner (i.e., termed ply-cord strike-through)
3. Filling potential void areas formed when the tire is shaped, which could occur in the shoulder region due to it being the area of greatest deformation during tire shaping. Contoured barrier/inner liner assemblies are designed to address this, thereby ensuring uniform gauge, bead to bead
4. In the case where a barrier is applied only in the shoulder region, the ply gauge can be adjusted to ensure adequate resistance to ply cord strike-through

The liner is a critical component in the modern radial pneumatic tire and, therefore, is the specific topic of Chapter 4.

3.6 PLY

The ply in many tires is typically produced from polyester, though, in some instances, nylon is used. In those cases, nylon can tend to show stress relaxation or creep. In the case of dual mounted tires, the expansion of the sidewalls of the adjacent tires can result in sidewall contact and consequent abrasion. As a result, polyester is preferred. An off-center ply is used when there is no barrier in the tire construction. The cords are located closer to the sidewall side of the ply, with more of the higher green strength barrier compound against the inside inner liner (Figure 3.10).

3.7 SIDEWALL

The sidewall typically extends from the exterior of the bead area to the shoulder of the tire, and has three functions:

1. Protects the tire from damage due to curbs or other foreign objects;

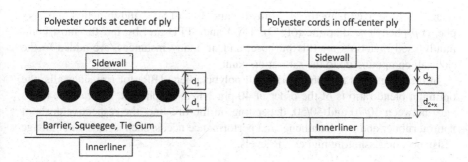

FIGURE 3.10   Off-center Ply Lay-up (6)

2. Resistance to fatigue and aging, as well as flexing, as the tire passes through its loaded footprint;
3. *Via* lettering, the sidewall provides structural information (plies, belts, load-carrying capacity) and other regulatory information, such as place and date of manufacture, and, if needed, white strips ("white sidewall"), raised white letters, the tire brand and name of the manufacturer.

Damage resistance is provided by two means: i) a high tear-strength compound, containing natural rubber and other compounding materials, to meet the performance and design requirement, and ii) if needed, a scuff rib which is a rubber strip extending outward around the circumference in the center region of the sidewall. Wear markers can also be molded into the scuff rib, thereby enabling the vehicle operator to know when the tire may need replacement due to sidewall abrasion. Such sidewall scuff ribs can be found on city bus tires which experience significant curb abrasion and potential damage.

Resistance to fatigue and aging is achieved by two techniques: (i) blends of natural rubber and polybutadiene, which are well known to allow improved resistance to fatigue and cut-growth (Figure 3.11), and (ii) use of antioxidants and antiozonants,

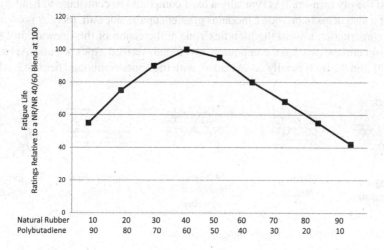

FIGURE 3.11   Natural Rubber and Polybutadiene Blends and Fatigue Life

such as polymerized trimethyl-1,2-dihydroquinoline (TMQ) and dimethyl butyl-N'-phenyl-*p*-phenylene diamine (6PPD). TMQ and 6PPD are, by far, the most commonly used materials for this purpose, and, in many instances, are added to the sidewall compound up to the saturation point.

In blending natural rubber (NR) and polybutadiene (BR) for tire sidewalls, the optimum blend ratio is of the order of 40 phr NR to 60 phr BR. In practice, ratios vary between 30/70 and 50/50, depending on the tire line, the respective costs of natural rubber and polybutadiene, and performance needs, such as sidewall damage resistance, necessitating higher NR levels.

3.8 BEAD

The bead serves to lock the tire onto the rim and help minimize the risk of rim slip, i.e., circumferential tire slippage on the rim, due to occasions of high drivetrain torque. The bead is produced by spiral winding of brass- or, sometimes, bronze-coated wire, previously coated with a compound called "bead insulation", to form a hoop of the specific dimension of the tire. Depending on the tire manufacturer, the bead can adopt different shapes (Figure 3.11). The more common types are a hexagonal configuration, tapered hexagonal, and a parallel piped rectangular configuration. Of the various types of beads, the tapered hexagonal and parallel piped rectangular configurations perform best in terms of locking the tire onto the rim and minimizing rim slippage.

Bias truck tires, with multiple fabric plies, require additional bead cables in order to anchor them. Figure 3.12 shows a bias truck tire with three bead cables, compared with a steel cord reinforced radial casing with only one bead.

3.9 APEX (BEAD FILLER)

The apex is the component above the bead, which fills the space between the inside ply and the ply turn-up. It is typically a hard compound to contribute to bead stability, but it also helps to provide a modulus gradient up the sidewall, thereby preventing stress construction toward the high flex zone at the center of the sidewall. In larger tires, sometimes two apex compounds are found, termed Apex #1 and Apex #2. Apex #1 and #2 are typically co-extruded, with the upper compound being #2, which

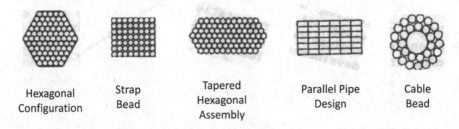

| Hexagonal Configuration | Strap Bead | Tapered Hexagonal Assembly | Parallel Pipe Design | Cable Bead |

FIGURE 3.12 Radial Tire Bead Configurations (4, 5)

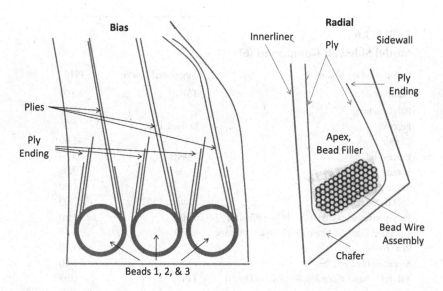

FIGURE 3.13 Bias versus Radial Tire Bead Areas [Bias construction with multiple plies (7)]

is also the softer of the two. Such configurations contribute to the required modulus gradient progression from the toe to the ply ending and then to the mid-section of the sidewall (Figure 3.13).

3.10 CHAFER (TOE GUARD)

This is a very hard compound at the interface between the tire and the rim. Its function is to prevent rim slippage, maintain dimensional stability of the lower bead region, and have sufficient durability to last the life of the tire.

3.11 SHOULDER WEDGE

Similar to the apex, this component fills the space underneath the belt edges, the ply, and the sidewall. It also serves to maintain a flat belt lay-up, ensuring, in turn, even footprint pressures. The compound is typically made of natural rubber and carbon black, and, though it has a high hardness (Shore A), it should be designed to ensure good component-to-component adhesion between the shoulder wedge and the sidewall, between the shoulder wedge and the ply, and between the shoulder wedge and the belts.

3.12 MODEL COMPOUND LINE-UP

Tables 3.6 to 3.12 are a collection of model starting point formulations for various components in a tire casing, such as the sidewall, inner liner, and ply compound.

TABLE 3.6
Model Sidewall Compound (6)

Compounding Material	Suggested Grade	PHR
Natural rubber	TSR10	50.00
Polybutadiene		50.00
Peptizer	Renacit 11	0.20
Carbon black	N330	45.00
Process oil	TDAE	3.50
Paraffinic wax		2.00
Microcrystalline wax		1.00
Tackifying resin	Escorez 1102	2.00
Polymerized trimethyl-1,2-dihydroquinoline	TMQ	1.50
Dimethyl butyl-N'-phenyl-p-phenylenediamine	6PPD	4.50
Zinc oxide		1.00
Stearic acid		1.00
t-Butyl-2-benzothiazolesulfenamide (TBBS)	TBBS	0.95
Sulfur		1.00
N-(Cyclohexylthio)phthalimide (CTP).	PVI	0.25
Total		163.90

TABLE 3.7
Model Inner Liner Compound (6)

Compounding Material	Suggested Grade	PHR
Bromobutyl rubber		100.00
Carbon black	N660	60.00
Process oil	Naphthenic	8.00
Homogenizing agent	Struktol 40 MS	7.00
Tackifying resin	Escorez 1102	4.00
Stearic acid		1.00
Mercaptobenzothiazole disulfide	MBTS	1.25
Zinc oxide		1.00
Sulfur		0.50
Total		182.75

3.13   SUMMARY

Regardless of the tire design and application, the construction of the casings is similar, consisting of a ply, anchored by the bead, and shaped by the constricting rigid belt composite. The goal of the engineer is to ensure that the casing meets the durability and safety standards required for the tire, which includes long-term aging resistance, adequate burst pressure resistance, and low hysteresis. The cooler the

TABLE 3.8
Model Shoulder Wedge Compound (6)

Compounding Material	Suggested Grade	PHR
Natural rubber	TSR20	100.00
Peptizer	Renacit 11	0.20
Carbon black	N550	45.00
Process oil	TDAE	3.50
Paraffinic wax		1.00
Microcrystalline wax		1.00
Tackifying resin	Escorez 1102	2.00
Polymerized trimethyl-1,2-dihydroquinoline	TMQ	1.00
Dimethyl butyl-N'-phenyl-p-phenylene diamine	6PPD	2.50
Zinc oxide		5.00
Stearic acid		2.00
t-Butyl-2-benzothiazolesulfenamide (TBBS)	TBBS	1.50
Sulfur		2.00
N-(Cyclohexylthio)phthalimide (CTP).	PVI	0.25
Total PHR		166.95

TABLE 3.9
Model Apex (Bead Filler) Compound (6)

Compounding Material	Suggested Grade	PHR
Natural rubber	TSR20	100.00
Peptizer	Renacit 11	0.20
Carbon black	N550	65.00
Process oil	TDAE	5.00
Paraffinic wax		1.00
Microcrystalline wax		1.00
Tackifying resin	Escorez 1102	4.00
Polymerized trimethyl-1,2-dihydroquinoline	TMQ	1.50
Dimethyl butyl-N'-phenyl-p-phenylenediamine	6PPD	1.50
Zinc oxide		5.00
Stearic acid		2.00
t-Butyl-2-benzothiazolesulfenamide (TBBS)	TBBS	1.50
Sulfur		2.50
N-(Cyclohexylthio)phthalimide (CTP).	PVI	0.25
Total PHR		190.45

TABLE 3.10
Model Wire Coat Compound (6)

Compounding Material	Suggested Grade	PHR
Natural rubber	RSS2	100.00
Peptizer		0.10
Carbon black	N325, N358	45.00
Silica	Conventional	10.00
Silane coupling agent	Si69	2.00
Process oil	TDAE	3.50
Paraffinic wax		1.00
Microcrystalline wax		1.00
Tackifying resin	Escorez 1102	2.00
Polymerized trimethyl-1,2-dihydroquinoline	TMQ	1.00
Dimethylbutyl-N'-phenyl-p-phenylene diamine	6PPD	2.50
Cobalt naphthenate		2.00
Zinc oxide		10.00
Stearic acid		2.00
t-Butyl-2-benzothiazolesulfenimide (TBSI)		0.70
Sulfur (insoluble)		8.00
N-(Cyclohexylthio)phthalimide (CTP).		0.25
Total PHR		191.05

TABLE 3.11
Model Barrier, Gum Strip and Cushion Compound (6)

Compounding Material	Suggested Grade	PHR
Natural rubber	TSR20	100.00
Peptizer		0.20
Carbon black	N326	45.00
Silica	Conventional	10.00
Silane coupling agent	Si69	2.00
Process oil	TDAE	3.50
Paraffinic wax		1.00
Microcrystalline wax		1.00
Tackifying resin	Escorez 1102	2.00
Polymerized trimethyl-1,2-dihydroquinoline	TMQ	1.00
Dimethyl butyl-N'-phenyl-p-phenylenediamine	6PPD	2.50
Zinc oxide		5.00
Stearic acid		1.50
t-Butyl-2-benzothiazolesulfenimide (TBSI)		0.70
Sulfur		1.50
Total PHR		176.80

TABLE 3.12

Model Passenger Tire Fabric Ply Compound (6)

Compounding Material	Suggested Grade	PHR
SIR 20	TSR20	70.00
SBR 1502		30.00
Peptizer		0.25
N330		50.00
Process Oil	TDAE	10.00
Resorcinol		1.00
TMQ		1.00
6PPD		1.00
Tackifying resin	Escorez 1102	4.00
HMMM		2.00
Stearic Acid		2.00
Zinc Oxide		4.00
TBSI		1.00
MBT		0.10
Insoluble sulfur (typically 25% oil extended)		3.00
PVI (CTP)		0.30
Total PHR		179.65

operating pressure of the tire, the better will be the long-term durability, and the tire will have directionally better rolling resistance.

All components in the casing play a critical role in meeting the needs of the tire design envelope. However, in terms of performance, the two most important components in the casing are the belts and the inner liner, and both components are the subject of more in-depth discussion in Chapters 4 and 5). Next would be the importance of applying all of the components in a uniform manner and will be covered with further discussion in Manufacturing (Chapter 8).

REFERENCES

1. Kovac F. *Tire Technology*, 5th Edition. The Goodyear Tire & Rubber Company, Akron, OH. 1978.
2. Rodgers B, D'Cruz B. Tire Engineering. Page 579 – 599. In *Rubber Compounding Chemistry and Applications*. Ed. B Rodgers. CRC Press, Boca Raton, FL. 2015.
3. Purdy JF. *Mathematics Underlying the Design of Pneumatic Tires*. Hiney Printing, Akron, OH. 1963.
4. Ford T, Charles F. *Heavy Duty Truck Tire Engineering*. Society of Automotive Engineers, SP-729, Warrendale, PA. 1988.
5. Rodgers B, Waddell W. The Science of Rubber Compounding. In *Science and Technology of Rubber*. Eds. JE Mark, B Erman, CM Roland. Academic Press, Boston, MA. 2015, 417–470.
6. Rodgers MB ELL Technologies LLC Personal communication. 2020.
7. Davison JA. *Design and Application of Commercial Type Tires*. Society of Automotive Engineers, SP-344, New York, NY. 1969.

4 Tire Innerliner

4.1 INTRODUCTION

There are two enabling technologies that have made possible the modern radial tubeless tire, namely i) the steel wire belts, which achieve a flat footprint, in turn, facilitating excellent wear and traction performance, and ii) the innerliner, to maintain tire inflation pressure without the use of a tube. The key enabler for the innerliner was the development of chlorobutyl rubber by Exxon Chemical Company, now ExxonMobil and based in the United States, and the development of bromobutyl rubber by Polymer Corporation of Canada, renamed Polysar in 1976, and now part of the Saudi Aramco corporation (1). Butyl rubber, invented by Standard Oil of New Jersey, now ExxonMobil, and its halogenated derivatives, chlorobutyl and bromobutyl rubber, have lower permeability properties compared with other elastomers, due to their tighter molecular packing and higher density, thereby reducing the flow rate of oxygen and nitrogen permeating through the tire innerliner, or the butyl inner tube in the case of tube-type tires (2). Gas diffusion through a polymer is very sensitive to polymer density and compound specific gravity, with small changes having a large effect.

There are three commercial halobutyl polymers: i) chlorobutyl was commercialized in 1961; then ii) bromobutyl in 1976, followed by a star-branched version of bromobutyl which mimics a bimodal molecular weight distribution for improved tire factory processing; and iii), a copolymer of isobutylene, p-methylstyrene and brominated p-methylstyrene, which was invented at ExxonMobil in the early 1990's and can be obtained under the commercial name, Exxpro™. Of the three types of halobutyl rubber, Exxpro™ is up to 20% better at maintaining the air retention of a tire. A model innerliner formulation using bromobutyl is illustrated in Figure 3.7. Though its air retention properties are the same as chlorobutyl, bromobutyl has emerged as the preferred polymer because of better innerliner compound adhesion to adjacent components in the tire (3).

The manufacture of the bromobutyl rubber is a two-step process. First, polymerization of isobutylene and isoprene takes place to produce butyl rubber (Figure 4.1). This is then followed by halogenation to form either chlorobutyl rubber, with chlorine, or bromobutyl rubber, with bromine.

4.2 PRODUCTION OF BROMOBUTYL RUBBER

Bromobutyl represents most of the global production of halobutyl rubber, and all of the new global capacity added in the past 10 years has occurred to produce this polymer. Though the permeability of butyl-, chlorobutyl-, and bromobutyl-based innerliners are comparable, bromobutyl allows a much higher level of adhesion of the innerliner to the tire casing barrier compound, which, in turn, is most important for long-term tire durability (3).

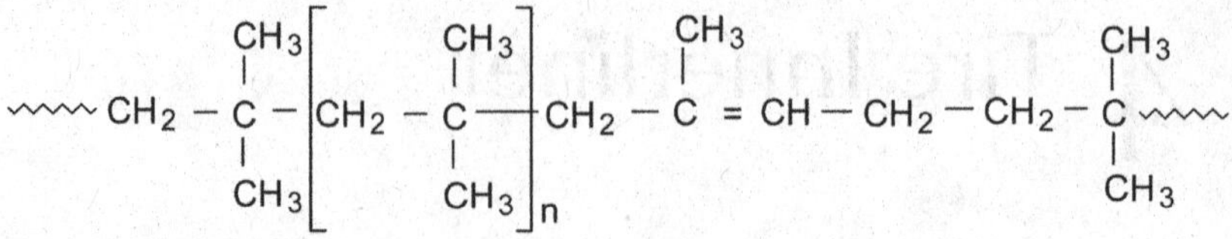

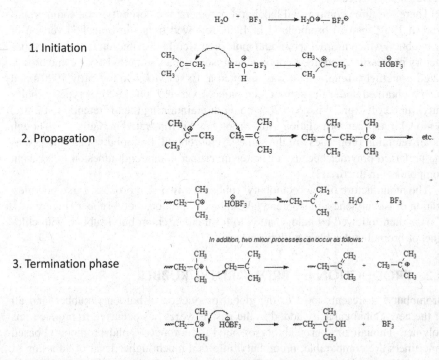

$$\sim\sim CH_2 - \underset{\underset{CH_3}{|}}{\overset{\overset{CH_3}{|}}{C}} - \left[CH_2 - \underset{\underset{CH_3}{|}}{\overset{\overset{CH_3}{|}}{C}} \right]_n CH_2 - \overset{\overset{CH_3}{|}}{C} = CH - CH_2 - CH_2 - \underset{\underset{CH_3}{|}}{\overset{\overset{CH_3}{|}}{C}} \sim\sim$$

FIGURE 4.1   Butyl Rubber, Copolymer of Isoprene and Isobutylene (4)

The first of two stages in the production of bromobutyl is the manufacture of butyl rubber from high-purity isobutylene and isoprene (5). In the polymerization processes, there are three steps: an initiation reaction, a polymer chain growth step or propagation, and thirdly, a termination or chain transfer step. In the case of butyl rubber, the mechanism of polymerization consists of complex cationic reactions. The catalyst system is a Lewis acid co-initiator with an initiator. Typical Lewis acids co-initiators used in commercial isobutylene polymerization include aluminum trichloride and alkyl aluminum dichloride. Brønsted acids, such as water, hydrochloric acid, organic acids, or alkyl halides, can be used as initiators. The isobutylene monomer reacts with the Lewis acid catalyst to produce a positively charged carbocation called a carbenium ion in the initiation step. On formation of the carbenium ion, isobutylene monomers are then added in the propagation step until chain transfer or termination reactions occur (Figure 4.2).

FIGURE 4.2   Simplified Overview of Butyl Rubber Polymerization (5) (Manufacturers can use different catalysts but the principles are similar)

Industrially, the most widely used production process for butyl rubber manufacturing is in a slurry, where particles of butyl rubber are formed in a methyl chloride suspension in the reactor after Lewis acid initiation. The reaction is highly exothermic, with the high molecular weight achieved by controlling the polymerization temperature, typically between −100°C and −90°C. Boiling liquid ethylene is used to remove the heat of reaction and maintain the required low temperature.

In addition to the low polymerization temperature, the final molecular weight of the butyl rubber also controls the initiation and chain transfer reaction rates. Water and oxygenated organic compounds, which can terminate the propagation step, are minimized by purifying the feed systems. After polymerization, methyl chloride and unreacted monomers are flashed off with steam and hot water, dried and purified, and then recycled back to the reactor. Stabilizers such as calcium stearate and an antioxidant are introduced to the hot water/polymer slurry to stabilize the polymer (1).

For regular butyl rubber production, the polymer is then removed from the hot water slurry, using screens, and dried in a series of extrusion, dewatering, and drying steps. Like in nearly all rubber manufacturing operations, fluidized bed conveyors are used to cool the product to a suitable packaging temperature, preferably below 30°C. Figure 4.3 illustrates the production process for butyl rubber.

Bromobutyl rubber is prepared by adding bromine (Br_2) to a hexane solution of butyl rubber at temperatures of around 50°C. Bromination occurs by an ionic substitution reaction at the isoprenoid units. The amount of bromine is controlled so that approximately one atom of bromine is associated with one isoprenoid unit of the

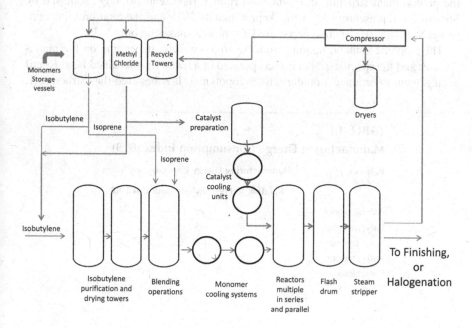

FIGURE 4.3 Manufacturing of Butyl Rubber (6)

elastomer. The hydrogen bromide generated in the reaction is neutralized by aqueous alkaline solution.

Calcium stearate, butylated hydroxytoluene (BHT), and epoxidized soybean oil (ESBO) are also added to prevent dehydrohalogenation and oxidation during finishing and storage. Briefly, calcium stearate is a mild base and reacts with excess hydrogen bromide (HBr), thereby serving as a stabilizer for bromobutyl rubber. It can also retard compound cure rates, so that the amount of calcium stearate used in bromobutyl rubber is tightly controlled. ESBO neutralizes any free acids forming in the rubber, such as stearic acid. BHT, which is used in the production of many elastomers, is an antioxidant that helps maintain polymer shelf life specifications and prevents degradative reactions.

The material is then dried in a series of dewatering steps, baled, and wrapped in plastic film. The manufacturing of butyl rubber is very energy intensive (Table 4.1). Polymerization occurs at −90 to −100°C, and, to finish the bromobutyl rubber, temperatures must be high enough to devolatilize moisture and hexane, which is the typical solvent used in the halogenation process (Figure 4.4).

The primary bromination reaction proceeds by the well-established bromonium ion mechanism, leading to a double bond shift and the formation of an exomethylene allylic bromide (Figure 4.5, Structure II). Limited allylic rearrangement and migration of the double bond can then follow.

The bromine content of bromobutyl rubber is expressed in weight percent (wt.%). Figure 4.6 shows the structures of the bromine-containing functional groups found in bromobutyl rubber. Total bromine content includes both organic bromine, as illustrated in Figure 4.6, and inorganic bromine, such as calcium bromide. Structure II, the predominant structure in bromobutyl rubber, represents 50–60%, followed by Structure I, representing 30–40%. Approximately 5–15% of the total bromine content is Structure III, whereas Structure IV typically makes up only 1–3%.

The isoprene content, ranging from 0.8 to 3.0 mole%, depending on the manufacturer and the specific grade, is incorporated in a *trans*-1,4 enchained head-to-tail arrangement to produce a random, liner copolymer. It is believed that Structure 1,

TABLE 4.1
Manufacturing Energy Consumption Index (8, 9)

Polymer	Manufacturing Energy Consumption Index Relative to Polyethylene (Index of 100)
Natural rubber	18
Polypropylene	95
Polyethylene	100
Polybutadiene	122
Polychloroprene	136
Emulsion SBR	100
EPDM	161
Solution SBR	147
Butyl rubber (IIR)	197

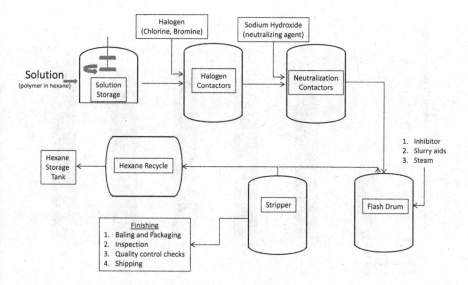

FIGURE 4.4 Halogenation (7)

$$\sim CH_2 - \underset{\underset{CH_3}{|}}{C} = CH - CH_2 \sim \xrightarrow{Br_2} \sim CH_2 - \underset{\underset{Br}{\overset{(+)}{|}}}{C} - \overset{(-)}{\underset{Br}{}} CH - CH_2 \sim \longrightarrow \sim CH_2 - \underset{\underset{Br}{|}}{\overset{CH_2}{\parallel}}{C} - CH - CH_2 \sim + HBr$$

Structure I   Structure II

FIGURE 4.5 Bromonium Ion Mechanism (1, 5)

$$-CH_2 - \underset{\underset{CH_3}{|}}{C} = CH - CH_2 -$$

Structure I

$$-CH_2 - \underset{\underset{Br}{|}}{\overset{CH_2}{\parallel}}{C} - CH - CH_2 -$$

Structure II

$$-CH_2 - \underset{\underset{CH_2Br}{|}}{C} = CH - CH_2 -$$

Structure III

$$-CH = \underset{\underset{Br}{|}}{\overset{CH_3}{}}{C} - CH - CH_2 -$$

Structure IV

FIGURE 4.6 Structure of Isoprenyl Units in Bromobutyl Rubber (10)

illustrated in Figure 4.5 and 4.6, is the dominant configuration, constituting around 94% of the enchained non-halogenated isoprene. Figure 4.7 illustrates the higher adhesion of inner liner to barrier compound possible with bromobutyl compounds, explaining its preference in modern radial tires.

4.3 TIRE INNERLINER

Air loss from a tire occurs *via* one of three routes, namely i) through the valve stem, ii) at the tire-to-rim interface typically caused by the unseating of a damaged rim

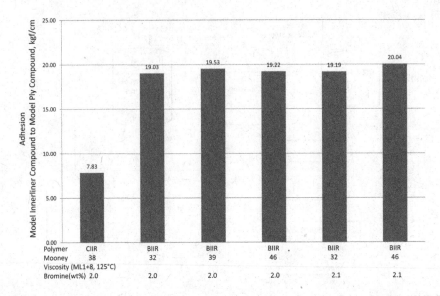

FIGURE 4.7 Adhesion of Chlorobutyl and Bromobutyl Innerliner Compounds to Tire Barrier Compound (3)

bead, due to severe maneuvering or rust, or iii) through the body of the tire. In turn, three factors control the air loss through the tire structure: the quality of the innerliner compound and its halobutyl content, the innerliner gauge, and the distance from the innerliner ending to the rim.

With regard to air loss attributable to the innerliner, this occurs *via* two paths, firstly, through the innerliner component, and secondly, through the bead area below the liner ending, if it does not extend fully to the toe area. Air then flows:

1. Through the permeable tire casing compounds
2. Along the ply cords, particularly if there are void areas within the cord (i.e., an open construction)
3. Along the belt wire void areas and belt compound
4. Through parts of the tire where gauges tend to be lower, such as in the mid sidewall and the base of the tread grooves.

Tire pressure generally decreases slowly through the innerliner over time. Gas is absorbed onto the innerliner surface and then slowly diffuses through the rubber from regions of high partial pressure, $p1$, to areas of low partial pressure, $p2$. This is a two-way process, from inside the tire pressure chamber to the outside atmosphere, and *vice versa*. Permeation (Q) through the liner is a function of the diffusivity (D) and the solubility parameters (h). It can be described by the empirical relationship in equation 4.1,

$$Q = D\,h\,A\left[\left\{p1 - p2\right\}\right]/dt \qquad (4.1)$$

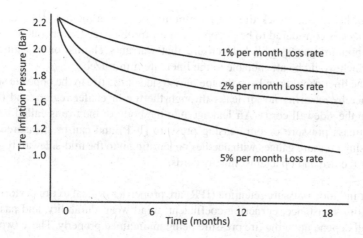

FIGURE 4.8 Innerliner Quality and Schematic of Tire Pressure Loss Rate

where A is the surface area of the innerliner, d is the innerliner gauge, and t is time. Figure 4.8 illustrates the impact of innerliner quality and IPLR on net tire inflation pressure with time.

The innerliner is a thin layer of rubber, applied as a sheet during tire building to the inside of a tubeless tire. Its primary function is air pressure retention. It is compounded to provide impermeability to gases, high flex fatigue and aging resistance, and adhesion to the casing compounds. The tire is built with the innerliner applied to the building drum as a flat rubber sheet, but, as the green tire composite is expanded during the shaping and curing processes, the ply cords can penetrate or deform the innerliner compound and *vice versa*. To minimize this distortion, the gauge is "contoured", i.e., the gauge across the green innerliner sheet varies according to high-stretch locations, when the tire is shaped. Also, high green strength barriers are used, thereby further reducing potential gauge distortion, in addition to preventing ply cords from cutting through the liner during tire building.

There is a set of empirical guidelines that can be considered in designing an innerliner component for a tire:

1. Compared to chlorobutyl innerliner compounds, bromobutyl shows much greater adhesion to the casing compounds. This improvement in component-to-component adhesion is due to more equivalent compound vulcanization rates, thereby minimizing the formation of an interface between the liner and adjacent components (11)

2. For chlorobutyl innerliner compounds to match the adhesion levels of the corresponding bromobutyl compounds, natural rubber is typically added to the formulation. This, however, has a negative effect on permeability, and, in turn, on tire inflation pressure loss rate (IPLR)

3. The optimum gauge for a passenger tire innerliner is 1.0 mm. This would ensure that the tire can meet an inflation pressure loss rate of below 2.5% per month which is the *de facto* General Motors and broad industry IPLR target

4. For heavy-duty truck tires, the optimum gauge is in the order of 2.0 mm, which is considered to be the optimum for trouble-free tire operation

5. Optimum gauges will help ensure that warranty claims, or adjustments which would be attributable to the liner, are minimized

6. The liner should run from the toe of one side of the tire to the opposite side, thereby ensuring no air leaks through the toe or chafer region, and then up the sidewall cords. Air leakage into the body of the tire is called intra-carcass pressure or intra-casing pressure (ICP) and can be measured by using pressure gauges with needles penetrating into the mid-sidewall region of the tire and up against the ply cords.

IPLR or inflation pressure retention (IPR) are properties critical to tire performance. Tire rolling resistance, cornering coefficient, tread wear, durability, and passenger safety all depend upon the tire pressure being maintained properly. These two terms merit further definition.

4.4   TIRE INFLATION PRESSURE LOSS RATE (IPLR)

The ability of a tire to retain air pressure can easily be measured, though tight temperature control is essential and atmospheric pressure must also be considered in making a determination. The IPLR is expressed in percent loss per month and the *de facto* industry target is 2.5% per month or lower. Air pressure retention of a radial tire is essential for its long-term performance and structural durability.

Air pressure loss and the tire's operation at low air pressure will result in structural weakness and degradation. Excessive oxidation will induce thermo-oxidative degradation of the casing compounds and belt materials, eventually causing a loss of adhesion, followed by separation. Ply end separations can occur due to increased flexing of the lower sidewall. whereas fatigue-related cracking of the innerliner and mid-sidewall can also cause tires to be removed from service.

The air retention characteristics of a tire are principally governed by the innerliner gauge, innerliner compound permeability, sometimes referred to as the coefficient, Q, and the innerliner ending in the bead region. Figure 4.9 shows how addition of natural rubber to the innerliner compound affects tire IPLR; the higher the natural rubber content, the greater the IPLR, and, in turn, the lower the ability of the tire to retain pressure.

4.5   TIRE INTRA-CARCASS PRESSURE (ICP)

As oxygen, nitrogen, and potentially water vapor penetrate through the innerliner due to the pressure differential between the inside of the tire and the external environment, gas pressure will build up inside the tire casing. As the tire operates, oxygen which has permeated into the casing can cause compound thermo-oxidative degradation. Bromobutyl innerliners provide the greatest resistance to such compound degradation, which can cause common belt edge separation and subsequent early tire removal. The higher the quality of the innerliner, i.e., the higher the halobutyl content of the innerliner, the lower will be the intra-carcass pressure or intra-casing

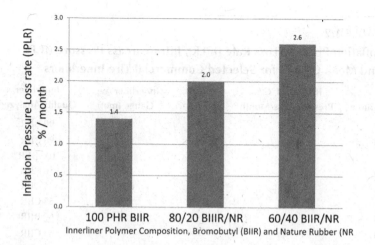

FIGURE 4.9 Innerliner Natural Rubber Content (PHR) and Inflation Pressure Loss Rate, IPLR (12)

pressure, in turn, lowering the potential for tire crown area belt edge separation and adhesion failure. In truck and bus radial tires, it also reduces the potential for rust of the steel tire ply cords.

The intra-carcass pressure (ICP) test measures the accumulated air pressure within the casing plies, and, as mentioned, is measured by attaching a hypodermic needle directly to a pressure gauge and inserting the needle through the mid-sidewall into the carcass plies. Large differences in ICP are found for tires, due, in most instances, to the quality of the innerliner compound (Figure 4.10). Tires with 100 phr

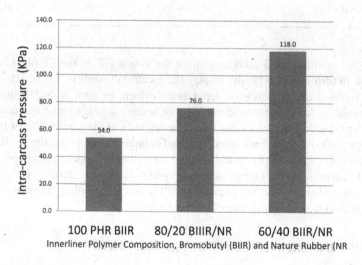

FIGURE 4.10 Innerliner Natural Rubber Content (PHR) and Intra-Carcass Pressure, ICP (12)

TABLE 4.2

Inflation Pressure Loss Rate (IPLR), Intra-carcass Pressure (ICP), and Mean Gauges for Selected Commercial Tire Innerliners

Source	IPLR at 21°C % Pressure Loss/Month	ICP at 21°C (Bar)	Innerliner Avg. Gauge (mm)	Innerliner Quality Analyzed
A	2.65	0.75	0.82	90 BIIR
B	1.55	0.6	1.07	90 BIIR
C	2.08	0.57	1.02	100 BIIR
D	4.2	1.22	1.49	30 CIIR
E	3.8	1.03	0.77	45 CIIR
F	3.52	0.95	1.08	55 CIIR
G	3.36	0.99	0.8	60 BIIR
H	3.44	–	0.77	55 CIIR

TABLE 4.3

Relationship between Halobutyl and Natural Rubber (IR) Content, Intra-Carcass Pressure, and Tire Durability

Rubber Hydrocarbon Content in Liner	HIIR/IIR, %Vol	Liner Permeability Comparative Rating at 65°C+	Equilibrium Intra-carcass Pressure, MPa	FMVSS 109 Step-Load Hours to Failure
100 BIIR	65.2	3	0.032	61.5
75 BIIR/25 NR	48.3	4.2	0.063	56.9
65 BIIR/25 NR/20 IIR*	42.2	5.9	0.063	40.2
60 SBR/40NR/20 IIR*	16.1	6.9	0.09	31.5

bromobutyl or chlorobutyl innerliners have the lowest ICP values (Table 4.2, Table 4.3), and, in turn, would typically display the greatest durability.

Tire removal due to belt edge separations, which, in turn, is due to innerliner performance, is due to compound oxidative or aerobic degradation. Anaerobic compound degradation can also occur, though much later in the service life of the tire. Such degradation mechanisms could be found in large truck and bus tires (TBR) and earthmover (OTR) tires. This could also be due to high tire operating temperatures. Figure 4.11 represents the impact innerliner quality has on tire durability; the higher the natural rubber content, the greater the IPLR and ICP, leading to greater oxidation, degradation, and finally loss of tire performance.

4.6 TIRE INNERLINER COMPOUND PROPERTIES

Examples of various grades of butyl and halobutyl rubbers, supplied by ExxonMobil, and which are considered to be of the highest quality available to the industry, are

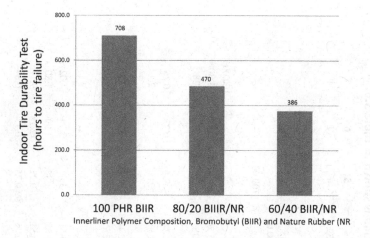

FIGURE 4.11 Impact of Increased IPLR and ICP on Tire Durability (13)

shown in Table 4.4. These polymers are known to process efficiently and have become recognized as the highest quality as defined by viscosity uniformity, purity, packaging, and supply reliability. Some properties of innerliner compounds using various grades of bromobutyl rubber from ExxonMobil are shown in Table 4.5. Air pressure retention of a radial tire is critical to its long-term performance and

TABLE 4.4

Examples of Halobutyl Rubber Supplied by ExxonMobil (14)

Butyl Polymer	ExxonMobil Grade	Mooney Viscosity (ML 1+8 at 125°C)	Isoprene (mol %)	Halogen	Halogen (mol %)	Typical Application
Butyl	065	32	1.05	–	–	Innertubes, bladders
	068	51	1.15	–	–	Innertubes, bladders
	268	51	1.70	–	–	Innertubes, bladders
	365	33	2.30	–	–	Innertubes, bladders
Chlorobutyl	1066	38	1.70	Cl	1.26	Innerliners, white sidewalls, innertubes
Bromobutyl	2222	32	1.70	Br	2.00	Passenger tire innerliner
	2235	39	1.70	Br	2.00	Passenger tire innerliner
	2255	46	1.70	Br	2.00	Truck tire innerliner
	2211	32	1.70	Br	2.00	Fast cure rate applications
	2244	46	1.70	Br	2.00	Fast cure rate applications
	6222	32	1.70	Br	2.00	Easy-processing innerliners

TABLE 4.5

Innerliner Compound Properties with ExxonMobil Bromobutyl (3) (Formulation is shown in Table 3.6)

Compound		1	2	3	4	5	6	7
Polymer Mooney viscosity		51	38	35	39	46	32	46
Butyl 268	[PHR]	100.00	–	–	–	–	–	–
CIIR 1066		–	100.00	–	–	–	–	–
BIIR 2222		–	–	100.00	–	–	–	–
BIIR 2235		–	–	–	100.00	–	–	–
BIIR 2255		–	–	–	–	100.00	–	–
BIIR 2211		–	–	–	–	–	100.00	–
BIIR 2244		–	–	–	–	–	–	100.00
Mooney viscosity (100°C)	[MU]	59.30	53.60	52.10	55.20	59.00	51.50	61.80
Monsanto Tel tack tester	[lb/in²]	23	25	23	25	26	25	28
Green strength								
Green strength – 100% strain	[N/mm²]	0.23	0.22	0.22	0.24	0.25	0.23	0.26
Stress relaxation – 75% decay	[sec]	122	101	99	170	218	97	200
Mooney scorch (125°C)								
Mm	[MU]	44.30	40.20	36.60	42.10	46.10	36.90	47.80
t5	[min]	–	15.90	25.93	22.73	20.58	24.15	19.92
Rheometer (MDR)		(160°C,	0.5° Arc)					
MH-ML	[dNm]	3.70	2.91	3.85	3.51	3.63	3.74	4.02
t10	[min]	7.23	1.12	2.03	1.82	1.83	1.99	1.80
t90	[min]	35.96	3.33	10.17	9.51	9.22	12.18	9.11
Peak rate	[dNm/min]	0.36	1.45	0.77	0.75	0.75	0.77	0.84
Rheometer (MDR)		(180°C	0.5° Arc)					
MH-ML	[dNm]	3.59	2.25	3.68	3.39	3.54	3.58	3.86

(Continued)

TABLE 4.5 (CONTINUED)
Innerliner Compound Properties with ExxonMobil Bromobutyl (3) (Formulation is shown in Table 3.6)

Compound		1	2	3	4	5	6	7
t10	[min]	1.87	0.54	0.89	0.80	0.83	0.85	0.81
t90	[min]	7.66	123	2.92	2.77	2.69	3.02	2.68
Peak rate	[dNm/min]	1.48	3.47	2.27	2.24	2.29	2.24	2.54
Tensile strength	[Mpa]	–	7.69	8.00	8.50	8.57	7.75	8.53
Elongation at break	[%]	1000.00	870	890	880	830	910	850
100% Modulus	[Mpa]	0.64	0.94	0.81	0.79	0.85	0.79	0.87
300% Modulus	[Mpa]	1.29	2.78	2.33	2.48	2.78	2.27	2.83
Hardness	[Shore A]	41	42	41	39	40	41	41
Tear strength	[K/mm]	3425	4029	38.11	41.26	41.75	3921	43.68
Adhesion to ply	[Kgf/cm]	–	7.03	19.03	19.50	19.55	19.19	20.04
Fatigue to failure								
Mean	[K-Cycles]	902	164.2	219.7	156.4	153.7	206.3	148.2
Max	[K-Cycles]	141.3	255.2	327.8	264.5	216.3	317.4	313.7
Mm	[K-Cycles]	66.1	124.2	124.5	422	95.4	91.7	72.1
Weighted average	[Cycles]	109791	200740	285044	223662	183889	262370	248901
MOCON O_2 permeability								
Test temp.	[°C]	40	40	40	40	40	40	40
Permeation	(cc-mm/m² day)	224.8	219.6	194.6	201.1	181.1	189.1	199.7
Permeability coefficient	(cc mm/m² day mm Hg)	0.332	0.325	0.288	0.297	0.268	0.280	0.296

structural durability. The determination of IPLR is a standard test method (14). Air pressure loss and/or operation at low air pressure result in the following structural weaknesses:

1. Excessive oxidation degradation of the carcass and belt rubbers, resulting in loss of adhesion and separation.
2. Break-above-bead failures due to excessive flexing of the lower sidewall.
3. Flex-cracking of the innerliner and mid-sidewall, eventually resulting in tire failure.
4. Separations due to excessive inter-component pressure build-up within the tire structure.

There are six properties which define the performance of a tire innerliner:

1. Compound viscosity
2. Adhesion to adjacent components in the tire
3. Permeability
4. Fatigue resistance
5. Green strength, i.e., the strength of the compound to resist component stretching and innerliner gauge distortion
6. Brittle point

All of these properties are largely determined by the halobutyl content of the compound. What merits a specific discussion on the use of halobutyl rubber in modern radial tubeless tire are the challenges in compounding, processing, and building tires with this polymer. Bromobutyl, chlorobutyl, and regular butyl rubber are classed as "non-compatible" polymers, and, if mixed with natural rubber, SBR, or polybutadiene compounds, or if such compounds become contaminated with butyl compound scrap, will result in reduced adhesion and tear strength. This is due to:

1. Large differences in solubility parameters from and non-miscibility with general-purpose elastomers, resulting in separate phase formation (Table 4.6)
2. Reduced crosslink density of butyl polymers *versus* general purpose rubbers (GPR)

TABLE 4.6
Polymer Solubility Parameters (16)

Polymer	Nominal Solubility Parameter δ (MPa ½)
Polybutadiene	17.2
Polyisoprene	16.4
SBR	17.1
Butyl, IIR, BIIR	16.1

3. Risk of contamination in wire coat compounds, with loss of adhesion
4. Contamination of tread compounds, with consequent loss in tear strength

Furthermore, of the range of different polymers used in a tire, halobutyl polymers have the highest molecular weights, necessitating their final storage in a "hot room" in the compound production area of the tire factory. This is necessary to optimize compound mix times and to aid dispersion of compounding ingredients. However, if the residence time in the "hot room" exceeds best practice, dehydrohalogenation, or evolution of the halogen, can occur, resulting in discoloration and in generation of scrap rubber. In addition to contamination concerns, which can have a detrimental impact on fatigue and crack resistance, other raw materials are used in halobutyl innerliners, that are seldom found in other tire compounds, include N660 carbon black, naphthenic or MES (Mild Extract Solvate) oil, and the accelerator, mercapto-benzothiazole disulfide (MBTS).

Air pressure retention has a direct impact on other tire performance parameters

- Rolling resistance of a tire is influenced directly by its inflation pressure, as a higher inflation pressure can result in a smaller tread footprint area and a lower drag force. It is estimated that running a tire at 20% under the recommended inflation pressure can increase fuel consumption by 10%. The Energy Department has estimated that, in the U.S., every psi of tire under-inflation wastes four million gallons of fuel daily.
- Tire cornering coefficient increases as inflation pressure increases. The road handling characteristics of the vehicle depend upon the inflation pressure of the tire being maintained at the recommended level. The inflation pressure determines the amount of flex in the tire sidewall.
- Tread wear is affected by inflation pressure. An optimized footprint is required for uniform wear, with an under-inflated tire exhibiting excessive wear on the shoulders and an over-inflated tire showing wear in the center. It is estimated that running a tire 25% under recommended inflation decreases tread life by about 25%.
- Tire durability is reduced when the proper inflation pressure is not maintained, as under-inflation generates excessive heat and stress on the tire which can reduce its service life.
- Finally, only when a tire is properly inflated can it perform as designed, providing the driver with the required tire–vehicle handling and steering control.

4.7   INNERLINER AND TIRE MANUFACTURING

The manufacture of tires involves a series of unit operations, consisting of compound mixing, component preparation, building, vulcanization or curing, final inspection, and, lastly, shipping (Figure 4.12). The first step is compound mixing, and a typical formulation is shown in Table 3.6. Innerliner compound represents around 6% of the tire weight. Briefly:

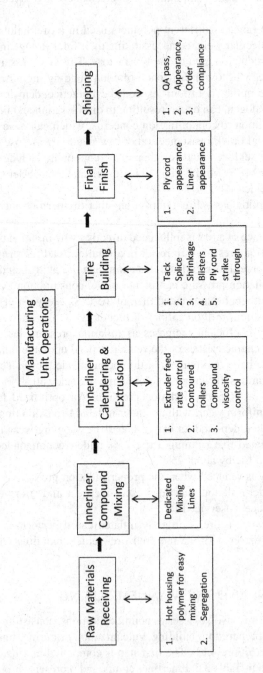

FIGURE 4.12 Innerliner Unit Operations and Processing Elements

TABLE 4.7
Target Mix Cycle Descriptions

Stage	Mix Time (Typical)	Target Drop Temperature	Maximum Drop Temperature
Non-productive	3 minutes	125°C–135°C	145°C
Productive	90 seconds	100°C	105°C

1. Innerliner compounds are typically mixed in two stages, a non-productive or masterbatch, and a final, or productive, stage. For definitions:
   a. Non-productive (masterbatch): This is where the polymer, carbon black or other fillers, processing aids, and oils are added
   b. Productive (final): The vulcanization system is added and includes sulfur, accelerators, and zinc oxide

2. This two-stage mix cycle compares favorably with the preparation of other compounds, such as natural rubber truck tread compounds, which could have four stages, silica treads with as many as six stages, and sidewalls and wire coat compounds with three stages (Table 4.7).

3. Drop temperatures and mix times for halobutyl innerliner compounds are shown in Table 4.7. These values compare with natural rubber or SBR compounds, where factory mix temperatures can be set as high as 180°C.

To minimize the risk of contamination, several options are available:

1. Tire factors with large enough production tonnages can dedicate a banbury mixer to mixing exclusively butyl or halobutyl compounds, thus minimizing the risk of accidental contamination with other compounds.

2. If two factories are in reasonable proximity to each other, one plant will dedicate capacity to mixing these compounds for both plants

3. One mixer is dedicated to mixing butyl compounds and, when mixing runs, for multiple days of tire production, are completed, thorough machine clean-out procedures are in place to remove any residual compound or other ingredients anywhere in the mixing line, before any other compound mixing is introduced (this can result in increased scrap, i.e., clean-out compound, and machine off-line time)

On completion of compound mixing, the innerliner is produced. The current trend is to extrude innerliners rather than calendar, due to the cost of calendar trains, labor, and throughput rates. The barrier, with this nomenclature varying among tire manufacturers, is best applied to the liner at the end of the extrusion line and before wind-up.

a. In many instances, the tie gum or barrier is passed through an electron beam radiation (EBR) unit before going to the innerliner extrusion line.

b. EBR processing results in increased green strength, allowing reduced gauge, and thus reduced cost.
c. Payback periods for EBR units are fast and of the order of months. This is due to the weight reduction for equivalent green-strength and materials savings.

Roller die extruders and calendars have contoured rolls to ensure that the final tire innerliner gauge is uniform. Heavy centerline gauges of innerliner sheet allow for more uniform gauge control in the final built tire after it has gone through the casing shaping turn-up. Friction ratios for halobutyl compounds are the reverse for general-purpose elastomers, e.g., natural rubber-based compound tie gum strips. Temperatures of the rolls of both extruders and calendars are similar, typically 75°C to 85°C. Output rates depend on the compound but are in the order of 6–7 m/min. Compound viscosity also depends on the formulation and target Mooney viscosities are shown in Table 4.8 This does not take into consideration control of viscosity and green strength for splice integrity. Caution is needed to control feed rates of extruder and depends on mills to prevent air being drawn into the compound later, to form blisters

The four primary issues pertaining to innerliners during tire building are: i) achieving a good splice, ii) adequate tack to ensure adhesion, iii) liner dimension stability, i.e., no innerliner sheet shrinkage over time, and iv) elimination of trapped air to prevent blisters. The build sequence is as follows,

1. The liner or liner–tie gum/barrier pre-assembly is typically the first compound to be applied at the building drum.
2. The component lay-down sequence is typically:
   a. Liner & barrier/tire gum
   b. Ply
   c. Flipper, chipper or other bead protection components, if used
   d. Bead assembly
   e. Sidewall and chafer/toe guard

The turn-up and shaping then follow, after which, in 2-stage building machines, the casing is moved to the 2nd station, where the belts and tread are applied.

The compound lay-down sequence may change with the design of the tire, construction, and configuration of the tire-building equipment, and the amount of

TABLE 4.8

Target Compound Viscosity with Process Equipment

Process Equipment	Target Mooney Viscosity Range (ML1+4, 100°C)
Cold feed extruder	60–65
Calendar (CFE)	65–75
Calendar (mills feed system)	70–75

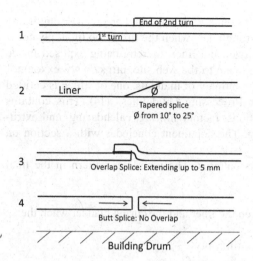

Splice and Component
Joint Configurations
1. Double wrap
2. Tapered
3. Overlap (up to 5 mm)
4. Butt: -][- with no overlap

FIGURE 4.13 Splice Configuration

component pre-assembly. The two most important factors to consider, regarding the liner in tire building, are:

1. Tack: if too high, the builder has difficulty in handling the liner sheet; if too low, it will not adhere to other tire parts during assembly
2. Splice: a low splice angle, thereby maximizing the splice overlap, is preferred. The greater the angle, the smaller the overlap, and the greater the potential for the splice opening during tire shaping (Figure 4.13). A double wrap of the liner on the building machine is preferred for TBR tires but, if possible, the ends should be tapered to avoid trapped air and subsequent cracking.

Tire inspection teams at the end of the manufacturing process check the innerliner for the following defects:

1. Compound degradation
2. Liner cracks
3. Thin spots
4. Blisters
5. Separations
6. Delamination
7. Liner ending adhesion
8. Splice integrity
9. Ply cord strike-through

Inspectors will pass the tire on to check for uniformity and then shipping, send it for further screening, such as balance or X-ray, or, if any defects are found, divert it for QA technical analysis.

High-viscosity halobutyl polymer compounds have higher green strength and are used to help resist cord strike-through, i.e., when the tire is too tight, ply cords can pull through the tie gum or barrier and liner, thereby being exposed inside the tire cavity. Reference should be made to the web site, https://www.exxonmobilchemical.com/. This web site has a number of manuals, one of which is entitled "Tire Halobutyl Rubber Innerliner Processing Guidelines" (14). This contains information on compound formulations, factory mixing, calendaring, and extrusion, and a section on tire building. The document concludes with a section on troubleshooting.

Tire innerliners can have various patterns. The innerliner pattern in the final cured tire has several purposes:

1. To help vent air trapped between the liner and the curing bladder when the tire is being vulcanized
2. To improve innerliner surface uniformity
3. To serve as a trademark for specific manufacturers

The pattern is created by the tire press curing bladder. When the tire is loaded, the press closes, and the bladder inflates at the start of the cure cycle. The pattern is then transferred from the curing bladder.

In conclusion, a check-list of suggested data to collect when running a tire innerliner compound mixing and processing trial should be prepared. Suggested data to be considered in the preparation of an innerliner could include:

1. Mooney viscosity (ML1+4 at 100°C),
2. Rheometer cure properties,
3. Shrinkage resistance,
4. Contoured innerliner dimensional stability,
5. Innerliner surface appearance.
6. Tensile strength elongation, and modulus (elongation at break can be an indicator of fatigue and crack resistance)

4.8   NITROGEN INFLATION AND NEW TECHNOLOGIES

In some instances, the original manufacturer of a vehicle will use nitrogen rather than air to inflate the new tires. Nitrogen inflation of tires is considered technically sound for a number of reasons. Approximately 21% of air is oxygen but, because oxygen has a smaller molecular diameter than nitrogen, it can migrate through a rubber barrier faster than nitrogen and subsequent pressure loss can thus be greater. Oxygen also supports combustion and can react chemically with rubber (oxidation). Since nitrogen is absorbed and penetrates through the liner at a slower rate than oxygen in air, inflation pressure maintenance is better and, in aircraft, high-performance racing and OTR tires, nitrogen reduces ignition risks. It is therefore normal practice for aircraft, racing, and large off-road tires to be inflated with nitrogen. For passenger car tires, however, in practice when inflating with nitrogen:

1. Multiple nitrogen purges still leave residual oxygen in the tire pressure cavity;
2. Atmospheric oxygen still permeates through tire components into the nitrogen-rich pressure tire chamber
3. Field tests by tire companies to date, comparing nitrogen with dry air inflation, have shown essentially equivalent performances, e.g. in terms of fuel economy and durability, and,
4. There is an extra expense and inconvenience to the use of nitrogen inflation.

For nitrogen inflation to be successful, further improvements in the quality of nitrogen gas (nitrogen produced using compressors can still contain up to 10% oxygen), and innerliner quality (i.e., 100 phr of bromobutyl) will be needed to ensure consumers see the full benefits.

Two other areas of potentially disruptive technological development involve replacement of the innerliner with a film, and the use of nanocomposites. Both areas are the subject of development, both in academia and at the tire companies, and, though both technologies face hurdles in their implementation, could offer the opportunity for further tire performance improvement.

4.9 VULCANIZATION OF HALOBUTYL TIRE INNERLINER COMPOUNDS

A final comment on the innerliner concerns the composition of the vulcanization system. In many instances, compounding of the innerliner, and specifically the cure system, has led to performance losses that could have been avoided. For example, because bromobutyl is a highly saturated polymer, engineers can compensate by adding higher levels of sulfur and accelerators, in an attempt to increase compound tensile strength and other mechanical properties. However, that could lead to loss of innerliner-to-casing adhesion. In order to ensure adequate adhesion and fatigue resistance, the industry today typically uses N660 carbon black, and the accelerator mercaptobenzothiazole disulfide (MBTS), with optimized levels of zinc oxide, stearic acid, and sulfur (Table 3.6).

It is essential for the engineer to understand the fundamental chemistry of the compounds used in a tire, primarily to minimize the risk of errors. With regard to innerliner vulcanization, two sets of reactions have been proposed, one based on carbon-carbon crosslink formation accelerated by zinc oxide, and the second, classical crosslinking by sulfur and MBTS, forming a sulfurating agent (11).

The crosslinking of bromobutyl rubber with zinc oxide involves the formation of carbon–carbon bonds through alkylation chemistry. The dehydrohalogenation of bromobutyl rubber to form a zinc halogen catalyst is a key feature of this crosslinking chemistry. During bromination of the isoprenoid unit of butyl rubber, an allylic bromide structure, involving an exomethylene group, appears as the predominant reactive structure. In the presence of zinc oxide, this structure undergoes initial crosslinking *via* bromine elimination to form allylic carbo-cations as shown in Figure 4.14.

In the unstabilized state, both the crosslinking rate and the halogen-elimination rate are many times faster for bromobutyl rubber than for chlorobutyl rubber. In the

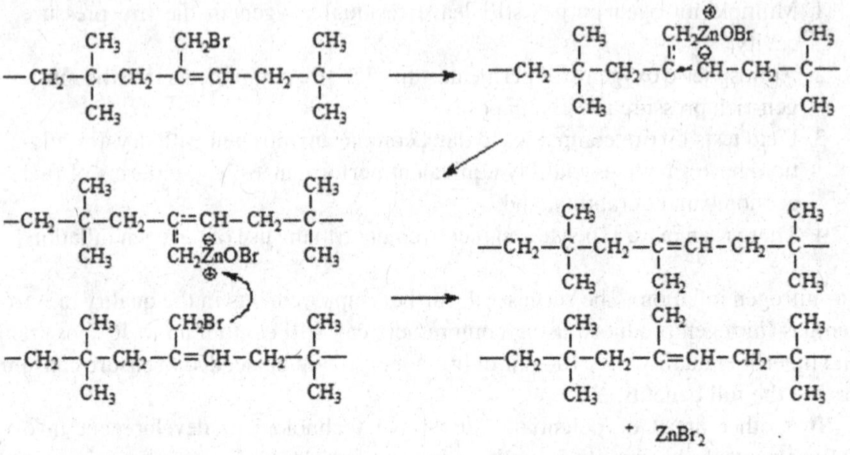

FIGURE 4.14 Formation of Allylic Carbo-cations during Vulcanization (11)

presence of a stabilizer, the differences between the rate constants for the two halo-butyl rubbers are substantially reduced. With the use of added retarders and more complex vulcanization systems, the apparent cure rate differences between bromo-butyl and chlorobutyl compounds may be adjusted still further.

4.10 SUMMARY

The tire innerliner is one of the most important components in a radial pneumatic tire and has a large influence on many performance parameters. All of the new global manufacturing capacity that has come on-line over the past 10 years has been for bromobutyl rubbers because: i) it offers better innerliner adhesion, and ii) air retention is better, because little to no natural rubber is required for adhesion, as is the case when using chlorobutyl innerliners

REFERENCES

1. Waddell WH, Tsou AH. Butyl Rubbers. In *Rubber Compounding, Chemistry and Applications.* Ed. B Rodgers. CRC Press, Boca Raton, FL. 2015, 109–138.
2. Boyd RH, Pant PVK. Molecular Packing and Diffusion in Polyisobutylene. *Macromolecules,* Vol. 24, pp. 6326–6331. 1991.
3. Wamsley E, Chen R, Rodgers B. *Designing Bromobutyl Elastomer Systems for Tire Innerliner Compounds.* Presented at a Meeting of the American Chemical Society, Rubber Division, Cleveland, OH. 2019.
4. Solis S, Rodgers B, Tambe N, Sharma, Waddell WH. *A Review of the Vulcanization of Isobutylene-Based Elastomers.* Presented at a Meeting of the American Chemical Society, Rubber Division, San Antonio, TX. 2005.
5. ExxonMobil Chemical Company. *Bromobutyl Rubber Compounding and Applications Manual.* BO917-05OE49. 2020. www.exxonmobilchemical.com/en/library.
6. Webb RN, Shaffer, TD, Tsou AH. *Commercial Isobutylene Polymers. Encyclopedia of Polymer Science and Technology.* Online Edition. John Wiley & Sons, New York, NY. 2003.

7. Rodgers MB. ELL Technologies LLC. Personal communication. 2020.
8. Gaines LL, Shen SY. *Energy and Materials Flow in the Production of Olefins and Their Derivatives*. Argonne National Laboratory (ANL). CNSV-9. U.S. Department of Energy. 1997. In-line; www.osti.gov.
9. *Energy Efficiency Index in Rubber Industry*. Thailand Ministry of Energy. www.dede .go.th.
10. Rodgers B, Solis S, Tambe N, Sharma BB. Alkylphenol Disulfide Polymer Accelerators and the Vulcanization of Isobutylene Elastomers. *Rubber Chemistry and Technology*, Vol. 81, pp. 600–624. 2008.
11. Rodgers B, D'Cruz B. *The Vulcanization System for Bromobutyl Based Tire Innerliner Compounds*. Presented at a Meeting of the American Chemical Society Rubber Division, Cleveland, OH. 2013.
12. ExxonMobil Chemical Company. *Exxon*™ Bromobutyl Rubber Compounding and Applications Manual. 2020. www.exxonmobil.com.
13. Waddell, W, Rodgers B. *Tire Applications of Elastomers. 3, Innerliner*. Paper No. J. Presented at a Meeting of the American Chemical Society, Rubber Division, Grand Rapids, MI. 2004.
14. ExxonMobil Chemical Company. *Exxon*™ Halobutyl Rubber Tire Innerliner Processing Guidelines. 2020. www.exxonmobil.com.
15. ASTM D3985. Standard Test Method for Oxygen Transmission Rate Through a Membrane or Film . *Annual Book of ASTM Standards*, Vol. 09, No. 01. 2017.
16. Grulke, EA. Solubility Parameter Values. In *Polymer Handbook*, 4th Edition, Eds. J Brandup, EH Immergut, EA Grulke. John Wiley & Sons, New York, NY. 1998, 675–711.

5 Tire Reinforcements

5.1 INTRODUCTION

Textile fibers and steel wire provide the reinforcement necessary for the tire to maintain its dimensions while under high inflation pressures and carrying the load at the vehicle- operating speed. For tire textiles, polyester is the dominant fiber in North America (55%), followed by nylon (43%) and then rayon. Globally, the distribution shifts, with nylon being the preferred reinforcement due to the higher proportion of bias tires, specifically, bias truck and bus tires, the favorable cost position of nylon, and its ready availability (1, 2, 3). The market is further segmented, according to segments and two markets: firstly, fabrics for bias and radial tires, and secondly, the products for the original equipment market and the larger, replacement market, driven by cost, availability, and performance.

The worldwide tire steel cord and tire fabrics industry is highly consolidated, with only a few major corporations dominating the market. These include Hyosung Corporation, Kolon Industries Incorporation, Indorama, SRF Limited, Bekaert, Kordsa Global Incorporation, Kordarna A.S., Cordenka GmbH & Co. KG, Teijin Limited, Tokusen USA Incorporation, and Milliken & Company, which are the key participants in the tire cord and tire fabrics industry. Table 5.1 gives a very brief summary of the history of the use of fabrics in tires where, today, hybrid cords are used which are made of both nylon and polyester yarns spun together to optimize conflicting needs for strength, fatigue resistance, and adhesion. The underlying trends among these manufacturers, driven by the market segmentation, is focused on three areas: i) new products such as aramids and hybrid cords, ii) growth (tonnage) to meet the global tire industry's quality requirements and supply security, and iii) consolidation to manage both supply and cost.

A tire is a textile–steel–rubber composite; the steel and textile cords reinforce the rubber and are the primary load-carrying structures within the tire. Because of the performance demands of fatigue resistance, structural rigidity, tensile strength, durability, and resilience, seven principal materials have been found to be suitable for tire application: cotton, rayon, nylon, polyester, steel, fiberglass, and aramid; the latter three materials find primary usage in the tire crown or belt region. Table 5.2 is a snapshot of the reinforcements currently used in the industry.

A number of fundamental tire properties, such as burst strength and casing dimensional stability, are dependent on the properties of the reinforcements (Table 5.3). Other properties, to varying degrees, are also affected by the reinforcement system, but, in all cases, they ultimately come down to tensile strength, fatigue resistance, and cord-to-rubber adhesion (Table 5.4).

The science of tire reinforcement employs a specialized terminology with which the tire engineer would be familiar:

TABLE 5.1
Summary of the History of Tire Reinforcement (4)

Introduction Period and Global Usage	Fabric	Treatment
Up to 1956	Cotton	Staple fiber, no treatment
1939 to present	Rayon	Stretched, then treated with RFL/ NR adhesive
1950 to present	Nylon	Stretched, then RFL/SBR or NR adhesive
1965 to present	Polyester	Stretch/heat set plus adhesive RFL/ blocked isocyanate
1970 to present	Steel cord of varying strengths	Initially copper coated, then brass
1975–1985	Fiberglass	Treated with RFL adhesive
1980 to present	Aramid	Treated with RFL adhesive
1995 to present	Hybrid polyester–nylon cords	Treated with RFL adhesive

TABLE 5.2
Tire Reinforcements for Different Tire Platforms (5)

Tire Line	Belts	Ply
Passenger	Steel	Polyester, hybrid cords
Radial light truck	Steel	Polyester, hybrid cords
Radial truck and bus	Steel	Steel
Bias truck tires	Polyester, nylon	Polyester
Farm	Polyester, rayon, nylon	Polyester, nylon
Off-the-road	Steel	Steel
Aircraft	Polyester	Polyester

TABLE 5.3
Reinforcement and Tire Properties (5)

Tire Performance Parameter	Effects of Reinforcement
Burst strength	Cord tensile strength
Bruise and cut resistance	Increase hot tensile strength and rigidity
Durability (separation resistance)	Fatigue resistance, adhesion
Power loss	Hysteresis, density
Tire dimensional stability	Modulus, creep, stress relaxation
Uniformity, flat spotting	Glass transition temperature, creep, modulus
Tread wear	Modulus
Noise	Creep
Spring rate	Modulus
Vehicle handling	Modulus

TABLE 5.4
Reinforcement Cord Properties

Mechanical Properties	Elements	Influencing Variable
Tensile strength	Modulus, elongation at break	
	Compression in rubber (buckling point)	Cord construction
Fatigue resistance	Stiffness	Cord elongation
Adhesion	Cord surface treatment	Open cord construction

1) Brass weight: Typically, 3.65 g/kg of cable; brass coat thickness is of the order of 0.3 µm.
2) Breaking strength: Tensile strength.
3) Cord: Structure consisting of two or more strands when used as a plied yarn or an end- product.
4) Denier: The weight of cord expressed in g per 9000 m.
5) EPI: Ends of cord per inch width of fabric.
6) Fibers: Linear macromolecules orientated along the length of the fiber axis.
7) Filaments: Smallest continuous element of textile, or steel, in a strand.
8) Filling: Light threads that run at right angles to the warp (also referred to as the "pick") that serves to hold the fabric together.
9) LASE: Load applied to a cord for a specified elongation (Load At Specified Elongation)
10) Ply twisting: Twisting of the tire yarn onto itself the required number of turns per inch; two or more spools of twisted yarn are then twisted again into a cord: for example, if two 840-denier nylon cords are twisted together, an 840/2 nylon cord construction is formed; if three 1300-denier polyester cords are twisted together, they give a 1300/3 cord construction.
11) Rivet: Distance between cords in a fabric; a high rivet count typically describes a fabric with a low EPI.
12) Tenacity: Cord strength, frequently expressed in g per denier.
13) Tex: Cord weight expressed in g per 1000 m.
14) Twist: Number of turns per unit length in a cord or yarn; direction of twist can be either clockwise ("S" twist) or counterclockwise ("Z" twist); twist imparts durability and fatigue resistance to the cord, though tensile strength can be reduced.
15) Warp: Cords in a tire fabric that run lengthwise.
16) Weft: Cords in a fabric running crosswise.
17) Yarn: Assembly of filaments.

Other important terms include:

1. Post-cure inflation: On completion of tire curing, it is immediately loaded onto a post-cure inflator which re-inflates the tire to 200 to 400 KPa (20 to 40 psi). This stretches the hot cords, thereby minimizing shrinkage, and

maintains the tire's shape and uniformity. Cooling is controlled to ensure a uniform temperature decrease to ambient conditions; afterward, the tire is released for final inspection

2. Flat spotting: A distortion or flattening of the tire, when loaded, occurring in the footprint after the vehicle has been parked in a warm environment. On starting up, the driver experiences the phenomenon as a severe vibration in the vehicle when driving.

3. Hybrid cords: Spinning together of nylon and polyester yarns, or other yarns, such as aramids, to optimize cord properties

5.2 FABRIC TIRE CORDS

Tire fabrics today are one of two primary types, nylon and polyester, and secondary types, namely rayon and aramids (4).

A. Rayon

Though its use is now very limited, rayon was the first synthetic fiber used as a tire reinforcement. Rayon is regenerated cellulose produced as a continuous filament and spun into a yarn. Because the polymer chains show strain crystallization, rayon has a high tensile strength, high dimensional stability, and good fatigue resistance. Though no longer used in North America because of its high cost and limited supply, it has found use in farm tires in Europe, where it is more available.

B. Nylon

Nylon is a polyamide polymer characterized by the presence of amide groups, -CO-NH-, in the main polymer chain. A wide variety of nylon polymers are available but only two have found application in tires: nylon 6,6 and nylon 6. Nylon 6,6 is produced from a condensation reaction between adipic acid and hexamethylenediamine. The "6,6" in the polymer designation denotes the six carbon atoms of hexamethylenediamine and the six carbon atoms of adipic acid. Nylon 6 is produced from caprolactam by a ring-opening polymerization. As caprolactam contains six carbon atoms and only one monomer is used, the polymer is thus designated "nylon 6". After the polymerization stage, the material is passed through a spinneret to form filaments, is cooled, and then twisted to form a yarn. This is then drawn out by up to 500% to orientate the polymer chains, create polymer crystallite zones, and increase the tensile strength.

Nylon finds use in the tire overlay or cap ply. When the new tire is cured, the nylon overlay will shrink, thereby locking the belts in place. Under high-speed tire service conditions, the belts show strong centrifugal forces but with a firmly positioned overlay, so that such forces are contained, the long-term durability of the tire is improved. The key properties for an overlay fabric are tensile strength and modulus, adhesion, shrinkage, and heat and fatigue resistance. For these reasons, nylon 6,6 is preferred. Depending on the speed rating of the tire, the overlay can take one of several lay-ups,

a) One layer or two layers
b) Covering the immediate top belt
c) Full belt coverage and partially extending over the shoulder wedge

In instances where nylon 6 is used, the tire will show "flat spotting", i.e., when the vehicle is parked, particularly in hotter environments, the loaded tire footprint will flatten. On start-up, the vehicle driver will experience severe vibration, due to this distorted section of the rotating tire. As the tire continues in service, the increase in operating temperature will allow the nylon 6 cords to relax and the temporary distortion in the tire may be mitigated. High-end tires tend to use overlays with a 1400×2 construction, whereas broad-market tires use overlay reinforcements, such as 840×2. This nomenclature, used for cord construction, is described below.

C. Polyester

Like nylon, a range of polyesters are available commercially for use in tires. Polyethylene terephthalate (PET) is the most important, though other polymers, such as polyethylene naphthalate (PEN), are emerging in importance, having lower shrinkage, higher modulus, and a 40°C higher glass transition temperature (120°C *versus* 80°C). Like nylon, polyester is formed by a condensation polymerization but with the monomers ethylene glycol and dimethyl terephthalate.

The polymerized material is then extruded through a spinneret to form filaments about 0.025 mm in diameter. These filaments are cooled, spun into a yarn, and drawn to give the required orientation and crystallinity to build strength. Polyester is the most common fiber used because it has low shrinkage, good fatigue resistance, and high tensile strength, and allows good adhesion build-up with rubber coat compounds. Polyester does not show the necessary shrinkage found with nylon, so it is typically not used in overlays.

D. Aramid

Another class of fibers which finds application in tires is known as the aramids. Kevlar is the trade name of the polymer, which has found its most extensive use in tires (Figure 5.1). Aramid is like nylon in that it contains the amide bond but is produced by copolymerizing terephthalic acid, used in polyester, and *p*-phenylenediamine. Hence, this aromatic polyamide is termed aramid (Table 5.5).

5.3 CORD CONSTRUCTION

Fundamentally, fibers are initially polymers. Molten polymer is passed through a spinneret and the resulting filaments are twisted into a yarn (Figure 5.2). The polymers tend to have higher molecular weights than those for applications other than in rubber products. To use the range of fibers for tire applications, the yarns must

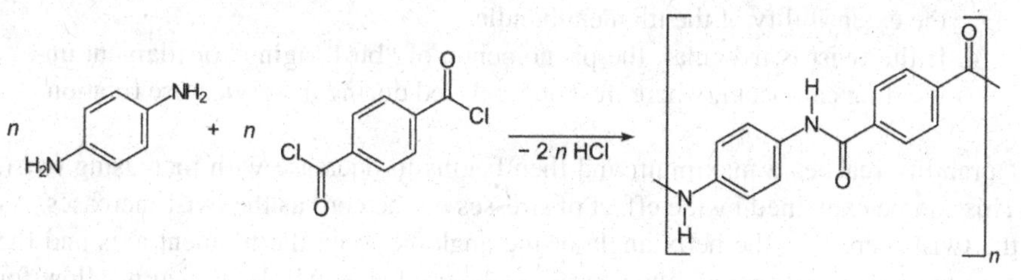

FIGURE 5.1 Aramid

TABLE 5.5
Fiber Applications in Tires

Fiber	Tenacity (g/ denier)	Elongation at Break (%)	Modulus (g/ denier)	Tire Application
Rayon	5.0	13	50	Farm
Polyester	6.5	18	65	Passenger ply
Nylon 6,6	9.0	19	32	Passenger, farm, aircraft
Nylon 6	9.0			Cost-sensitive agriculture tire applications
Aramid	20.1	4	350	Hybrid cords
Steel	3.8	2.5	200	Truck ply, all tire line belts

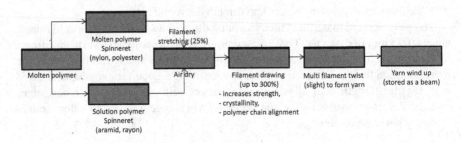

FIGURE 5.2 Schematic for Tire Cord Yarn Production

be twisted and processed into cords. First, yarn is twisted on itself to give a defined number of turns per inch, i.e., ply twisting. Two or more spools of twisted yarn are then twisted into a cord. Generally, the direction of the twist is opposite to that of the yarn; this is termed a balanced twist. There are a number of reasons for twist in a tire cord;

1. Twist imparts durability and fatigue resistance to the cord, though tensile strength can be reduced.
2. Without twist, the compressive forces would cause the cord outer filaments to buckle.
3. Increasing twist in a cord further reduces filament buckling by increasing the extensibility of the filament bundle.
4. If the twist is irregular, the phenomenon of "bird caging" or filament unwinding can occur, where the cord is flexed during in-service tire rotation

Durability reaches a maximum and then begins to decrease with increasing twist. This can be explained by the effect of stresses on the cord as the twist increases. As the twist increases, the helix angle or the angle between the filament axis and the cord axis, increases. In addition to twist, the cord size may be varied to allow for different strengths, depending on the application or tire line.

Generally, three-ply cords have the greatest durability. After cable twisting, the cords are woven into a fabric, using small fill threads. These threads are also referred to as picks. This weaving process introduces an additional construction variable, i.e., the number of cords per inch or EPI (ends per inch) that are woven into the fabric. High-end-count fabric gives greater plunger strength or penetration resistance. Low-end-count fabrics have more rivet (distance between cords) and give better separation resistance because of the greater rubber penetration around the cords. In addition, the weight savings may enable reductions in rolling resistance with equivalent tire strength factors.

5.4 FABRIC PRODUCTION

The most critical stage in preparing a cord or fabric for use in tires is fabric treatment, which consists of applying an adhesive under controlled conditions of time, temperature, and tension. This process gives the fabric the following properties:

1. Adhesion for bonding to rubber
2. Optimization of the physical properties of strength, durability, growth, and elongation of the cord for tire application
3. Stabilization of the fabric
4. Equalization of differences resulting from the source of supply of the fiber

Processing consists of passing the fabric through a series of zones which can be viewed as follows:

1. Adhesive application zone or first dip zone
2. First drying zone
3. First heat treatment zone
4. Second dip zone
5. Second drying zone and then second heat treatment zone
6. Final cooling zone

To obtain optimum cord properties of strength, growth, shrinkage, and modulus, specific temperatures and tensions are set at various exposure times within the fabric processing unit. The temperature and tensions determine, in part, the ratio of crystalline and amorphous areas within the fiber, and the orientation of the crystallites, which, in turn, determines the physical properties of the cord. For example, polyester, when heated, tends to revert to its un-orientated form and the cord shrinks. Stretching the cord in the first heating zone and then allowing the cord to relax in a controlled manner in the second heat treatment zone, i.e., stretch relaxation, will control shrinkage.

Another variable is change in processing temperature. An increase in temperatures can decrease cord tensile strength and modulus but will improve fatigue life which may be necessary for some tire constructions. However, not all cord properties behave similarly with changes in processing conditions. It is thus necessary to determine the processing conditions that optimize the specific cord properties needed for the required tire end use. When two or more diametrically opposed properties have to be optimized, more complex processing operations could be required.

5.5 CORD-TO-RUBBER COMPOUND ADHESIVE

There are three aspects to adhesion of tire cord to the elastomer treatment: molecular, chemical, and mechanical. Molecular bonding is due to absorption of adhesive chemicals from the adhesive dip or elastomer coating onto the fiber surface by diffusion and could be achieved by hydrogen bonding and van der Waals forces. Chemical bonding is achieved through chemical reactions between the adhesive and the fabric and rubber, i.e., crosslinking and resin network formation. Mechanical adhesion is a function of the quality of coverage of the cord by the rubber coating compound; the greater the coverage, the better the adhesion. The fiber properties of primary importance to adhesion are reactivity, surface characteristics, and finish. Rayon has many reactive hydroxyl groups. Nylon is less reactive but contains highly polar amide linkages, whereas polyester is quite inert. Thus, an adhesive system must be designed for each type of fiber. Regardless of the fiber, each adhesive system must conform to a rigid set of requirements:

1. Rapid rate of adhesion formation
2. Compatibility with many types of compounds
3. No adverse effect on cord properties
4. Heat resistance
5. Aging resistance
6. Good tack
7. Mechanical stability

The adhesive bond between the rubber and cord is achieved during the tire vulcanization cycle. The rate of adhesive formation should give maximum adhesion at the point of pressure release in the cure cycle.

5.6 STEEL CORD

Steel wires used in tires are of various configurations, but all are brass-coated wire strands, wrapped together to give cords of different characteristics, depending on the application. Steel tire cord is manufactured from high-carbon-steel rod, which is first drawn down to a diameter of approximately 1.2 mm. A brass plating is then added to the wire before a final drawing to 0.15–0.40 mm. These filaments are next stranded to form a cord construction, which is designed and optimized for a specific service requirement (4, 6, 7).

Because of the performance demands to which the tires are subjected, steel tire cord must be manufactured from high-quality steel. The composition of a typical steel cord is illustrated in Table 5.6.

One of the many trends in steel cord technology has been to increase its tensile strength. A series of steel wires, described as normal-tensile, high-tensile, super-tensile, and ultra-tensile strength steels, have emerged (Table 5.7).

In essence, as the strength of the steel increases, filament and cord strengths increase so that, for a given tire load-carrying capacity, steel cord diameters can decrease, thereby reducing tire weights and, in some instances, achieving improvements in rolling resistance.

The key mechanical properties governing a steel cord or wire are its tensile strength, elongation, and bending stiffness. A tire cord construction is normally defined by the structure, the length of lay, and the direction of lay (Figure 5.3).

TABLE 5.6
Nominal Steel Wire Composition; Normal Strength, NT (6)

Element	Amount (%)	Purpose
Carbon	0.70	Strength
Manganese	0.60	De-oxidation
Silicon	0.25	De-oxidation
Sulfur	0.11	Machinability
Phosphorus	0.12	Strength
Chromium	0.05	Strength
Cooper	0.02	Strength
Nickel	0.02	Strength

TABLE 5.7
Steel Cord Tensile Strength (2)

Steel Type and Strength	Carbon Content	Nominal Tensile Strength (MPa)
Normal tensile strength (NT)	0.7	2850
High-tensile (HT)	0.8	3300
Super-tensile (ST)	0.90	3650
Ultra-tensile (UT)	0.95	4000

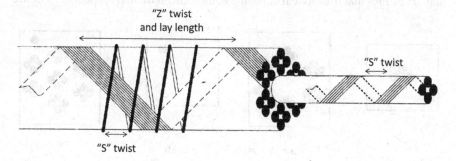

"Z" twist
and lay length

"S" twist

"S" twist

FIGURE 5.3   Length of Lay and Direction of Lay (Figure 5.4 for Description)

The full description of a steel cord is given by;

$$(N \times F) \times D + (N \times F) \times D + (F \times D) \tag{5.1}$$

Where N = number of strands, F = number of filaments, and D = nominal diameter of filaments (in mm) (Figure 5.4). An example of a steel cord specification would therefore take the form

$$(1 \times 4) \times 0.175 + (6 \times 4) \times 0.175 + (1 \times 0.15) \tag{5.2}$$

When N or F equals 1, the nomenclature system allows their exclusion. Thus, Equation 5.2 is reduced to

$$4 \times 0.175 + (6 \times 4) \times 0.175 + 0.15 \tag{5.3}$$

A number of additional conventions are used in defining a steel tire cord:

1. If the diameter D is the same for two or more parts in a sequence, then the diameter is specified only at the end of the sequence.
2. The diameter of the spiral wrap is specified separately.
3. When the innermost strand or wire is identical to the adjacent strand or wire, the definition of the wire can be further simplified by specifying only the sum of the identical components. Then, Equation 5.3 becomes

$$7 \times 4 \times 0.175 + 0.15 \tag{5.4}$$

4. The sequence or order in a wire designation follows the sequence of manufacture. The length and direction of lay for the cord in Equation 5.4 can be described as illustrated in Figure 5.3.

A number of empirical guidelines govern the construction of a wire for use in tires. For example, if the wire is used in a ply rather than in belts, it will undergo a greater amount of flexing. Hence fatigue performance will be important. If the application is in belts, then stiffness becomes a primary design parameter. Thus, key design properties can be specified and have been briefly summarized in Table 5.8.

The thicker a cord is, the stiffer it will be. Thinner cords tend to show better fatigue resistance. Heavier cords tend to find use in the larger tires, such as heavy-duty truck tires and tires for earthmoving equipment. Wire finds applications in tire

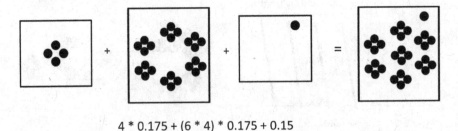

4 * 0.175 + (6 * 4) * 0.175 + 0.15

FIGURE 5.4   Wire Construction Illustrated in Figure 5.3 (4, 6)

reinforcement in tire belts, heavy-duty tire plies (e.g., in large truck tires), beads, and chippers (which protects the bead from wheel rim damage).

Figure 5.5 further illustrates how cord construction is described, with the number of filaments and the filament gauge. The third construction has a wire wrap intended to prevent the cord from unravelling or becoming loose. Table 5.9 shows some typical commercial wire constructions and their properties.

TABLE 5.8
Tire Cord Design Variables

Parameter	Design Variable
Stiffness	Cord filaments, gauge
Strength	Gauge, filaments, steel composition
Fatigue	Filament gauge
Elongation	Twist, lay length

Cord design example
-interior filament gauge 0.20 mm
-exterior wire gauge 0.38 mm

3x 0.20 + 6x 0.38

Cord design example
-all filament gauges 0.22 mm

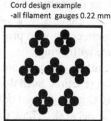

7 x 4 x 0.22

Cord design example
-interior filament gauge 0.22 mm
-exterior wire gauge 0.22 mm
-wrap wire gauge 0.15 mm

3 + 9 x 0.22 + 1x0.15

FIGURE 5.5   Illustration of Wire Construction and Description

TABLE 5.9
Examples of Commercially Available Wire Constructions and Nomenclature

Construction Type	Lay Length (mm)	Lay Direction	Cord Diameter	Breaking Strength (N)
Regular tensile strength cord				
2+1×0.25	14	S	0.63	375
2+2×0.25	14	S	0.66	490
3×3×0.15	9, 8	SZ	0.63	400
3+9×0.175+0.15	5/10/3.5	SSZ	1.00	755
3+9×0.22+0.15	6.3/12.5/3.5	SSZ	1.17	1210
High elongation cord				
4×4×0.22 HE	—	—	1.35	1150
3×7×0.22 HE	—	—	1.52	1720
High tensile strength cord				
2×0.30 HT	14	S	0.60	405
3×0.20+6×0.35 HT	10/18	SZ	1.13	1770
12×0.22+0.15 HT	12.5/3.5	SZ	1.18	13.6

5.7   MECHANISM OF BRASS-COATED STEEL WIRE-TO-RUBBER ADHESION

Much research continues on the mechanism of brass- and bronze-coated steel cord to rubber compounds. However, in the context of this discussion, the author had earlier prepared a summary of the mechanism and important compounding materials used in building sufficient adhesion for the required tire durability performance (4). A brief summary of the earlier work is important to ensure that the tire engineer avoids the risk of errors being incorporated in tire specifications. The thin coating of brass on the steel cord is the primary adhesive used in steel-to-rubber bonding. The quality of this bonding system, built up during vulcanization of, say, a radial tire, will influence the performance of the steel ply or steel belt in the tire and, ultimately, affect the durability of the product. Though the mechanism of bond formation in rubber–steel cord adhesion is very complex, in this case, a brief review of the current understanding of wire-to-rubber adhesion is pertinent. Natural rubber, typically used in wire coat compounds, forms a strong bond with brass as a result of the formation of an interfacial copper sulfide (CuS) film during vulcanization. In the case of radial tire wire coat compounds, when general-purpose rubbers, such as polybutadiene or SBR, were used, compound properties, such as fatigue and cut growth resistance, tend to decrease with aging.

In building adhesion, copper sulfide domains are created on the surface of the brass film during the vulcanization reaction. Such domains have a high specific surface area and grow within the wire coat compound before the viscous polymer phase is crosslinked into an elastomeric network. Thus, the polymer molecules become locked into the crystalline copper sulfide lattice (Figure 5.6).

Important factors governing this bonding are the formation of copper sulfide, cohesive strength, adhesion to the brass substrate, and the rate of secondary corrosion reactions underneath the copper sulfide film. Zinc sulfide and iron sulfide do not bond because they do not grow rapidly enough during vulcanization, do not form porous domains, and thus cannot interlock with the polymer. As the primary requirement is the formation of a copper sulfide domain before the initiation of crosslinking, the subsequent reduction of compound scorch time can adversely affect bond formation.

Mechanical stability of the copper sulfide domains is essential to retain the long-term durability of the rubber-to-wire adhesion. However, corrosion of the

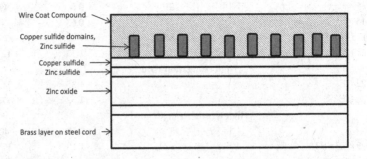

FIGURE 5.6   Copper Sulfide in Rubber-to-Brass-Coated Wire Adhesion

wire–rubber adhesive bond is catalyzed by Zn^{2+} ions that diffuse through the interfacial CuS layer. This will eventually result in an excess of either ZnS or $ZnO/Zn(OH)$. Under dry conditions, this process is slow. Nevertheless, Zn^{2+} will migrate to the surface with a consequent drop in mechanical interlocking of the CuS domains and rubber, followed by adhesion loss. Migration of Zn^{2+} ions is a function of the electrical conductivity of the brass coating. Addition of Co^{2+} or Ni^{2+} ions will reduce this conductivity.

Cobalt salts in the wire coat compound act to accelerate the vulcanization rate of high-sulfur compounds, which wire coat compounds tend to be. There are four types of cobalt salts used in tire wire coat compounds,

1. Cobalt stearate
2. Cobalt naphthenate
3. Cobalt neodecanoate
4. Manobond ™ (cobalt boron complex, e.g., 680C)

The Manobond group of cobalt salts are particularly effective at maintaining long-term wire-to-rubber adhesion, due to other additives included with the cobalt salt when it is produced by the supplier (Figure 5.7).

The increase in crosslink density increases the pull-out force of the wire in the rubber. More importantly, cobalt salts form Co^{2+} ions at the interface of the brass surface during vulcanization, and this will affect copper sulfide formation. Differences in efficiencies between cobalt adhesion promoters are due to the ease with which Co^{2+} ions can be formed. For example, zinc or brass reacts more easily with cobalt boron decanoate complexes than with cobalt naphthenate or cobalt stearate. The Co^{2+} and Co^{3+} ions are incorporated into the ZnO film before the sulfide film has been built up. Both di- and trivalent cobalt ions reduce the electrical conductivity of the ZnO lattice, thereby reducing the diffusion of Zn^{2+} ions through the semiconducting film. Diffusion of metallic copper domains to the surface following oxidation by R-S_x is not affected, as Cu^{2+} ions migrate along grain boundaries of the ZnO layer. Thus, if a cobalt salt is used, formation of copper sulfide at the cord surface will be accelerated, whereas the generation of ZnS will be hindered (Figure 5.8).

In rubber-to-brass-coated wire, a second parameter is important and that is the physical or radial pull-out force required to remove a wire. A high pull-out force is important in long-term adhesion retention (8). It is typically achieved *via* use of a reinforcing resin system, such as resorcinol and hexamethoxy methyl melamine (HMMM) (Figure 5.9). In some instances, this would therefore be used only in bronze-coated wire. With resorcinol, HMMM will react at vulcanization temperatures to form an interpenetrating network (IPN), resulting in an increase in stiffness (G′) and tensile

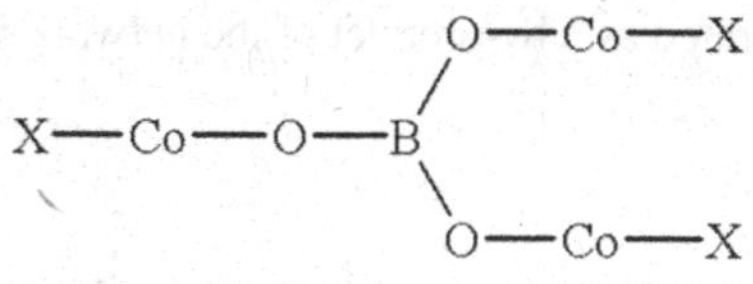

FIGURE 5.7 Manobond™ Cobalt Salts

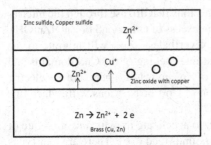
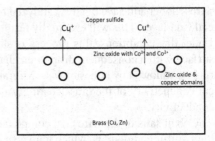

FIGURE 5.8 Cobalt Presence and Copper Sulfide Formation. Cobalt Reduces Brittle Zinc Sulfide by Hindering Zinc Ion Migration (4)

FIGURE 5.9 HMMM and Resorcinol IPN Formation

strength (9). This review is necessarily brief, and the reader is encouraged to consult additional references for further detail on the chemistry of rubber-brass adhesion (4). Hexamethylenetetramine (HMTA) may also be used in some instances, but, in the case of brass-coated wire, it would tend to oxidize the steel cord, possibly as a result of the ammonia being generated as a by-product of the network formation (Figure 5.10).

5.8 BEAD WIRE

Tire beads are built from single steel wires or heavy-gauge filaments. The typical composition of bead wire is described in Table 5.10. The tire bead is produced by

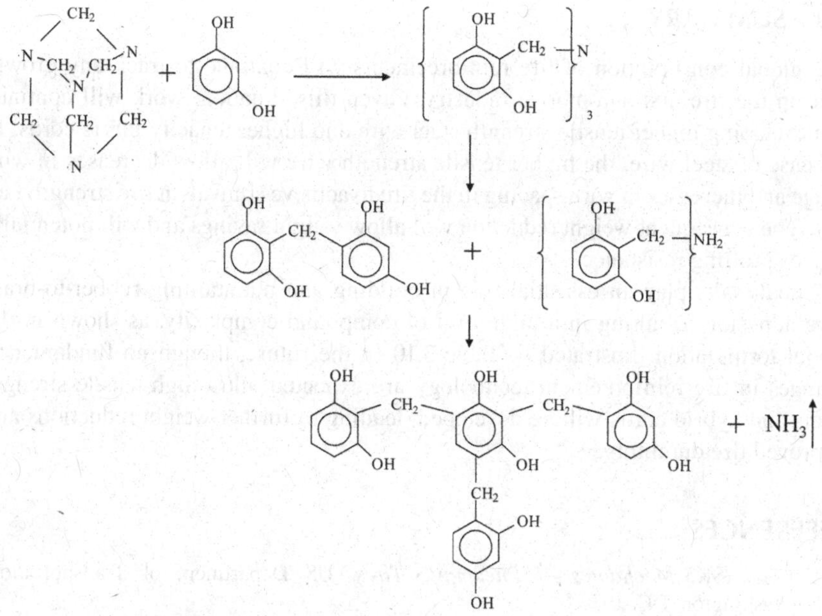

FIGURE 5.10   HMTA and Resorcinol IPN Formation

winding the wire on a drum, building the assembly up to the specific diameter for the bead of the tire in which it will be used. Examples of different bead configurations have been illustrated in Figure 3.11. The wire is available in both rectangular and round forms. In bead building, the wire is first coated with an adhesive compound, called the bead insulation, and the operation is in-line with the bead winder. The bead insulation is, in many instances, based on emulsion SBR or it can be based on an in-line tie gum natural rubber formulation. The bead wire can be bronze- or brass-coated, thereby offering greater flexibility in the design of the adhesive system contained in the bead insulation compound.

TABLE 5.10
Composition of Bead Wire (6, 7)

Tensile Strength	Normal (NT)	High (HT)
Carbon	0.725	0.825
Manganese	0.550	0.520
Silicon	0.250	0.230
Sulfur	0.150	0.008
Phosphorus	0.150	0.010
Copper, chromium, nickel	Trace	Trace

5.9 SUMMARY

The global consumption of tire reinforcements will continue to track any growth seen in the tire and automotive industry. Given this situation, work will continue on developing higher tensile strength steel wire and higher tenacity fabric cords. In the case of steel wire, the higher tensile strength wire will allow decreases in wire gauge and increases in wire spacing in the tire to achieve equivalent tire strength factors. The consequent weight reduction will allow weight savings and will potentially improve rolling resistance.

Cobalt salts play an essential role in building and maintaining rubber-to-brass wire adhesion, resulting in a high level of compound complexity, as shown in the model formulation illustrated in Table 3.10 In the future, though no fundamental changes in tire reinforcement technology are expected, ultra-high-tensile strength steels and hybrid cords will be developed, leading to further weight reductions and improved tire durability.

REFERENCES

1. Clark SK. *Mechanics of Pneumatic Tires.* US Department of Transportation, Washington, DC. 1984.
2. Gent AN, Walter JD. *The Pneumatic Tire.* US Department of Transportation, National Highway Transportation and Safety Administration, Washington, DC. 2005.
3. Shimar A. *Tire Cord Fabrics Market Global Forecast to 2023.* Markets and Markets, Northbrook, IL 2019.
4. Rodgers B, Waddell W. Tire Engineering. In *Science and Technology of Rubber.* Eds. J Mark, B Erman, C Roland. Academic Press, NY. 2013, 653–694.
5. Kovac, F. *Tire Technology.* The Goodyear Tire & Rubber Company, Akron, OH. 1978.
6. *Bekaert Steel Cord Catalogue.* N.V. Bekaert S.A. B8550 Zwevegem, Belgium. 1991.
7. www.bekaert.com. 2019.
8. Rodgers B, Jacob S, Curry C, Sharma BB. *Butyl Rubber Curing Bladder Resin Vulcanization Systems: Compositions and Optimization.* Presented at a Meeting of the American Chemical Society Rubber Division, Pittsburgh, PA. 2009.
9. Rodgers, B, Halasa A. *Compounding and Processing of Rubber/Rubber Blends. Encyclopedia of Polymer Blends.* Vol. 2, pp. 109–162. Processing 1st Edition. New York. 2011.

6 Radial Tire Materials Technology and Rubber Compounding

6.1 INTRODUCTION

Along with design and construction, tire material technology is one of the three core elements in tire engineering. For tire applications, compounded rubber has many unique characteristics not found with other materials, such as dampening properties and hysteresis, high elasticity, and abrasion resistance. Compounding is a highly complex technology, involving many traditional disciplines such as organic chemistry, polymer chemistry, materials physics, mathematics, and engineering mechanics. The raw materials available to the tire engineer for formulating a rubber compound can be divided into five categories:

1. Polymers including natural rubber and synthetic elastomers, and more importantly, blends of elastomers
2. Filler systems, for example, carbon black, clays, silicas, and calcium carbonate
3. Stabilizer or protectant systems: antioxidants, antiozonants, and waxes
4. Components of the vulcanization system: sulfur, activators, and accelerators
5. Special materials: secondary components such as pigments, oils, resins, processing aids, and short fibers

The use of each class of these materials in tires is reviewed in this chapter.

6.2 POLYMERS USED IN TIRES

Total global rubber production in 2019 is of the order of 28.8 million tons of which 13.8 million tons (48%) is natural rubber, with synthetic rubber making up the balance of 15.0 million tons (52%). Synthetic elastomers, in turn, are termed either General-Purpose Rubbers (GPR), such as styrene butadiene rubber (SBR) and polybutadiene (BR), or Special-Purpose Rubbers, such as butyl rubbers. Natural rubber, considered to be a general-purpose elastomer, is obtained mostly from Southeast Asia or Africa. Global output of natural rubber is dependent on several factors, such as market demand, pricing, new plantations, and replacement yield rates, and, to a lesser extent, weather patterns. In addition, despite well over 75% of natural rubber coming from smallholdings, supply lines have been very stable. Synthetic rubbers are produced from monomers obtained from the cracking and refining of petroleum (1, 2).

Though all elastomers play an essential role in the performance of a new radial tire, natural rubber usage has increased more than other elastomers. Between 1975 and 2008, natural rubber's share of the total rubber market has increased. There have also been diverse government incentives to help develop the downstream rubber processing industry in Southeast Asia. Though the majority of production is still exported, natural rubber consumption has still trended upward in these countries,. The increased rubber consumption in Southeast Asia has therefore been due to increasing consumption in the natural rubber-producing countries as well as the greatly increasing demand for natural rubber in China and India. The fundamental reasons for the increase in consumption have been attributed to improved green strength, better component-to-component adhesion, improved tear strength, lower tire temperatures generated under loaded dynamic service conditions, and lower tire rolling resistance to improve vehicle fuel efficiency which are all necessary requirements for modern radial tires (Figure 6.1). The increase in natural rubber usage translates into approximately 21 kg per tire for a radial truck tire compared with approximately 9 kg found in a bias truck tire (1) (Table 6.1).

Natural rubber (NR) is among the most important elastomers used in the manufacturing of tires. Roberts has reviewed natural rubber, covering topics ranging from basic chemistry and physics, to production and applications, and Rodgers

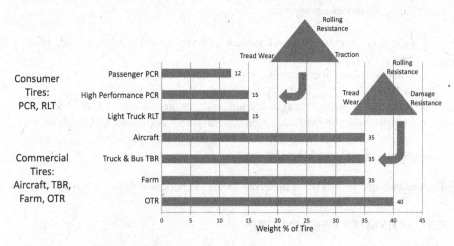

FIGURE 6.1 Natural Rubber Content in Radial Tires, Illustrating the Strategic Importance of the Material in Tire Engineering (1)

TABLE 6.1
Natural Rubber Use in Bias and Radial Tires (1)

Natural Rubber %	Bias Tires	Radial Tires
Tread	47	82
Coat compounds	70	100
Sidewall	43	58

has published comprehensive overviews of natural rubber and compounding (1, 3). This material, which represents a truly renewable resource, comes primarily from Indonesia, Malaysia, India, and the Philippines, with Southeast Asia representing nearly 90% of the total global acreage devoted to natural rubber plantations. Many more additional sources of good-quality natural rubber are becoming available and this is expected to continue to develop as favorable pricing and demand continue to increase (Figures 6.2 and 6.3). It is a material that is capable of rapid deformation and recovery, and it is insoluble in a range of solvents, though it will swell when immersed in organic solvents at elevated temperatures. Some of its many attributes include abrasion resistance, good hysteretic properties, high tear strength, high

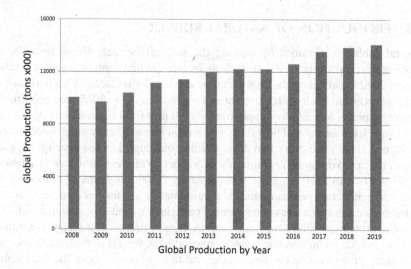

FIGURE 6.2 Global Natural Rubber Production Growth, Driven Largely by Radial Tire Production Increases (2, 4)

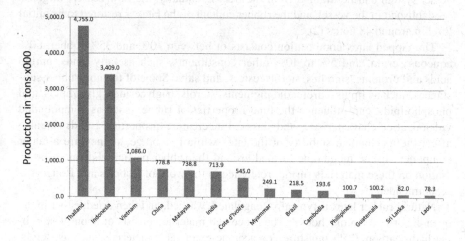

FIGURE 6.3 Global Natural Rubber Production Sources [2018] (4)

tensile strength, and high green strength. However, due to strain crystallization, it may also display poor fatigue resistance. It may be difficult to process in some factories, and it can show poor tire performance in areas such as traction or wet skid, when compared with synthetic elastomers, such as SBR.

Given the central role of natural rubber in radial tire engineering, three elements in the technology of natural rubber are of importance:

* Production of natural rubber
* Industry classification, descriptions, and specifications
* Fundamental technological properties of natural rubber

6.3   PRODUCTION OF NATURAL RUBBER

Natural rubber is obtained by tapping the side of the tree, *Hevea brasiliensis*. Tapping starts when the tree is from 5- to 7-years-old and continues until it reaches around 20–25 years of age, after which time it is usually replaced. A knife is used to make a downward cut from left to right and at around a 20- to 30-degree angle to the horizontal plane, to a depth of approximately 1.0 mm from the cambium. Latex can then exude from the cut and flow from the incision into a collecting cup. Rubber trees are tapped about once every two days, yielding one cupful of latex each time, each containing approximately 50 grams of solid rubber. With trees cultivated at a density of 375 per hectare (150 per acre), approximately 2,500 kilograms of rubber can be produced per hectare per year which is approximately one ton per acre per year.

Rubber occurs in the trees in the form of particles suspended in serum, which also contains proteins, amino acids, and various carbohydrates. The serum constitutes the latex, which, in turn, is contained in specific latex vessels in the tree. Latex constitutes the protoplasm of the latex vessel and tapping or cutting of the latex vessel creates a hydrostatic pressure gradient along the vessel, with consequent flow of latex through the cut. In this way, a portion of the contents of the interconnected latex vessel system can be drained from the tree. Eventually, the flow ceases, turgor is reestablished in the vessel, and the rubber content of the latex is restored to its initial level in around 48 hours (2).

The tapped latex composition consists of between 30% and 35% rubber, 60% aqueous serum, and 5% to 10% other constituents, such as fatty acids, amino acids and proteins, starches, sterols, esters, and salts. Some of the non-rubber substances, such as lipids, carotenoid pigments, sterols, triglycerides, glycolipids, and phospholipids, can influence the final properties of rubber, such as compounded vulcanization characteristics and classical mechanical properties. Lipids can also affect the mechanical stability of the latex while in storage, as they are a major component in the membrane formed around the rubber particle. For more information on these materials further reference to the work of Roberts and Rodgers is recommended (1, 4).

Natural rubber latex is typically coagulated, washed, and then dried, either in the open air or in a 'smoke house'. The processed material consists of about 93% rubber hydrocarbon, 0.5% moisture, 3% acetone-extractable materials, such as sterols, esters, and fatty acids, 3% proteins, and 0.5% ash. Raw natural rubber gel content

TABLE 6.2
Definitions of Natural Rubber Terms

Cup lump	Bacterially coagulated polymer in the collection cup
Earth scrap	Collecting vessel overflow material collected from the tree base
Hevea brasiliensis	Natural rubber-producing tree
ISNR	Indian Standard Natural Rubber
Lace	Trim from the edge of collecting vessels or cups and coagulated residue left around the bark of the tree, where the cut has been made for tapping
Latex	Fluid in the tree obtained by tapping or cutting the tree at a 22° angle to allow the flow into a collecting cup.
LRP	Large rubber particles
Ribbed smoked sheets	Abbreviated to RSS and is produced from whole field latex sheets
Serum	Aqueous component of latex which consists of lower molecular weight materials such as terpenes, fatty acids, proteins, and sterols.
SMR	Standard Malaysian rubber
SRP	Serum rubber particles
TSR	Technically specified rubber
Whole field	Fresh latex collected from trees

can range from 5% to 30%. High gel content can create processing problems in tire or industrial rubber product factories. Nitrogen content is typically in the range of 0.3% to 0.6%. For further clarity, Table 6.2 presents a number of definitions of terms used in natural rubber production and supply (1)

The rubber from a tapped tree is collected in three forms: latex, cup-lump, or lace. These are collected as follows:

1. Latex, collected in cups, is coagulated with formic acid, crumbed or sheeted. The sheeted coagulum can be immediately crumbed, aged, and then crumbed or smoke-dried at around 60°C to produce typically ribbed smoked sheet (RSS) rubber. Air-dried rubber is finished in the form of air-dried sheets (ADS).
2. Cup-lump is produced when the latex is left uncollected and, due to bacterial action, is allowed to coagulate on the side of the collecting cup. Field coagulum, or cup-lump, is eventually gathered, cut, cleaned, creped, and crumbed. Crumb rubber can be dried at temperatures up to 100°C.
3. Lace is the coagulated residue left around the bark of the tree where the cut has been made for tapping. The formation of lace seals the latex vessels and stops the flow of rubber latex. It would typically be processed with cup-lump.

The natural rubber-processing factories obtain the raw material collected from trees in large plantations and also from smaller, independent holdings, in one of two forms, namely field coagula or field latex. Field coagula consist of cup lump from the collection cups, and tree and cup lace are obtained, for example, from the rim of the cup. The lower grades of material are prepared from cup-lump, partially dried smallholder's rubber, rubber tree lace, and earth scrap after cleaning. Iron-free water is necessary to prevent polymer oxidation. Field coagula and latex are the base raw materials for the broad range of natural rubber grades.

6.4   NATURAL RUBBER PRODUCTS AND GRADES

Natural rubber used in tires is available in three basic forms:

1. Sheets, both smoked and air dried
2. Technically specified rubber
3. Crepe, sheet rubber with a more defined technical description, and latex concentrates and specialty rubbers, which have been mechanically or chemically modified.

Of these three broad groups, the first two represent up to 90% of the total natural rubber produced in the world. For tire manufacturing, ribbed smoked sheets (RSS) and technically specified rubbers are the most important. Crepe rubbers are now of minor significance in the world market, accounting for less than 75,000 tons per year. Field coagulum grade block rubbers have essentially replaced brown crepes except in India. Only Sri Lanka and India continue to produce latex crepes. Figure 6.3 presents a simplified schematic of the process followed in the production of natural rubber.

RSS rubbers are made from intentionally coagulated whole field latex. They are classified by a visual evaluation. To establish acceptable grades for commercial purposes, the International Rubber Quality and Packing Conference have prepared a description for grading, with details given in the Green Book (5). Whole field latex is used to produce ribbed smoked sheet and is first diluted to 15% solids, and then coagulated for around 16 hours with dilute formic acid. The coagulated material is then milled, water is removed, and the material is sheeted over a rough surface to facilitate drying. Sheets are then suspended on poles for drying in a smokehouse for one to seven days. Only deliberately coagulated rubber latex, processed into rubber sheets, properly dried, and smoked, can be used in making RSS.

A number of prohibitions are also applicable to the RSS grades. Wet, bleached and rubber that is not completely dry at the time of the buyer's visual inspection are not acceptable (except for slightly under-cured rubber as specified for RSS5). Skim rubber, made of skim latex, cannot be used in whole or in part in the patches as required under packing specifications. Prior to grading RSS, the sheets are separated, inspected and any blemishes are removed by manually cutting and removing defective material.

Fresh *Hevea* latex has a pH ranging from 6.5 to 7.0 and a density of 0.98 g/cm3. The traditional preservative is ammonia, which, as a concentrated aqueous solution, is added in small quantities to the latex when collected from the cup. Tetramethylthiuram disulfide (TMTD) and zinc oxide are also used as preservatives due to their greater effectiveness as a bactericide. Most latex concentrates are produced to meet the International Standard ISO 2004 (6). This standard defines

TABLE 6.3
Ribbed Smoked Sheet Classification (1)

RSS Grade	Rubber Mold	Wrapping Mold	Opaque Spots	Over Smoked Spots	Oxidized Spots	Burnt Sheet	Comments
1x	No	No	No	No	No	No	Clean, dry, no blemish
1	Very Slight	Very Slight	No	No	No	No	Clean, dry, no blemish
2	Slight	Slight	No	No	No	No	No foreign material
3	Slight	Slight	Slight	No	No	No	No foreign material
4	Slight	Slight	Slight	Slight	No	No	No foreign material
5	Slight	Slight	Slight	Slight	N/A	No	N/A

the minimum content for total solids, dry rubber content, non-rubber solids, and alkalinity (as NH_3).

Table 6.3 provides a summary of the criteria assessed by inspectors in grading RSS. In essence, the darker the rubber, the lower the grade rating. In practice, the premium grade is RSS1, with the lower-quality grade being typically RSS4.

Air-dried sheets are prepared under conditions very similar to those for smoked sheets but dried in a shed without smoke or additives, with the exception of sodium bisulfate. Such rubber therefore lacks the protection against oxidation afforded by drying the rubber in a smokehouse. Regarding the use of this material, it can be substituted for RSS1 or RSS2 grades.

The International Standards Organization (ISO) first published a technical specification (ISO 2000) for natural rubber in 1964 (6, 7). Based on these specifications, Malaysia introduced a national Standard Malaysian Rubber (SMR) scheme in 1965, and, since then, all the natural rubber-producing countries started production and marketing of technically specified rubbers based on the ISO 2000 scheme. Technically specified rubbers are shipped in 'Blocks' which are generally 33.3 kg bales in the international market and 25.0 kg bales in India (Figure 6.4). All block rubbers are also guaranteed to conform to certain technical specifications as defined by the national schemes or by ISO 2000 (Table 6.4).

6.4.1   TSR-L

This is a light-colored rubber produced from high-quality latex, with low ash and dirt content. TSR-L is packed and presented in the same way as TSR-CV. The benefit of TSR-L is its light color along with its cleanliness and superior heat-aging resistance. TSR-L shows high tensile strength, modulus, and ultimate elongation at break properties for both black and non-black mixes.

This natural rubber grade can be used for light-colored and transparent products such as surgical/pressure-sensitive tape, textiles, rubber bands, hot water bottles, surgical and pharmaceutical products, large industrial rollers for the paper printing industry, sportswear, bicycle tubes, chewing gum, cable covers, gaskets, and adhesive solutions and tapes.

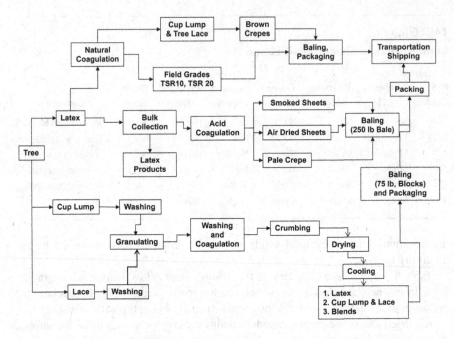

FIGURE 6.4   Production of Natural Rubber (1)

TABLE 6.4
Technically Specified Rubber (1)

Grade	Units	TSR CV	TSR L	TSR S	TSR 10	TSR 20
Dirt content	wt%, maximum	0.05	0.05	0.05	0.1	0.2
Ash content	wt%, maximum	0.6	0.6	0.5	0.75	1
Nitrogen content	wt%, maximum	0.6	0.6	0.5	0.6	0.6
Volatile matter	wt%, maximum	0.8	0.8	0.8	0.8	0.8
Initial Wallace plasticity	Po, minimum		30	30	30	30
Plasticity retention index	Minutes	60	60	60	50	40
Color (maximum)	Lovibond units		6			
Mooney viscosity	ML1+4 (100°C)	60+/−5				

6.4.2   TSR-5

This is produced from fresh coagulum, air-dried sheets, or RSS that have not been through the smoking process. It is packed and shipped to the same specification as that of TSR-CV and TSR-L. TSR-5 is typically used for general-purpose friction and extruded products, small components in passenger vehicles, such as resilient bush/ mountings, sealing rings, cushion gum, wire skim compounds in tires, and brake seals. Non-automotive applications include bridge bearings, ebonite battery plate, separators, and adhesives.

6.4.3 TSR-10

TSR-10 is produced from clean and fresh field coagulum or from sheets that have not been through the smoking process. It is packed and shipped in the same way as TSR-CV, TSR-L, and TSR-5. TSR-10 has good technological properties, similar to those of RSS-2 and RSS-3, but has an advantage over RSS in terms of:

- lower viscosity
- easier mixing characteristics (more rapid breakdown)
- technically specified and packed in 33.3 kg bales.

It can be used for tires, cushion gum stocks, joint rings by injection molding, micro-cellular sheets, for upholstery and packing, conveyor belts, and footwear.

6.4.4 TSR-20

In terms of quantities produced, this is a very important grade of technically specified natural rubber. It is produced mostly from field coagulum, lower grades of RSS, and unsmoked sheets. It is packed and shipped to the same specification as that of TSR-CV, TSR-L, TSR-5, and -10. TSR-20 has good processing characteristics and physical properties. Its low viscosity and easier mixing characteristics (compared with the RSS grades) can considerably reduce the mastication and mixing period. It is used mostly for tires, cushion gum stock, bicycle tires, micro-cellular sheet for upholstery and packing, conveyor belts, footwear, and other general products.

6.4.5 TSR-50

This is the lowest grade of TSR and is produced from old dry field coagulum or partly degraded rubber. It is packed and shipped in the same way as that of other grades of TSR.

It should be noted that these specifications will continue to be improved as production methods improve. For example, in 1991, the Rubber Research Institute of Malaysia revised the dirt levels of SMR CV60, CV50, and L from 0.05 to 0.025 weight percent, or wt%, SMR 10 from 0.10 to 0.08 wt%, and SMR 20 to 0.016 wt%. In addition, Malaysia has produced grades of rubber outside the specific scope of the ISO 2000 specification. SMR GP is a standard general-purpose (GP) rubber made from a 60:40 mixture of latex-grade sheet rubber and field coagulum. It is viscosity stabilized at 65 Mooney units, using hydroxylamine neutral sulfite (HNS). It is similar to SMR 10 in specification (3).

Table 6.5 presents a generalized quality ranking of the major grades of natural rubber (1). To illustrate consumption, distribution, and use of these various grades, shipments of SMR from Malaysia are typically: SMR 20, 60%; SMR 10, 25%; SMR CV & L, 5%; SMR GP, 8%; and SMR 5, 2.0%.

TABLE 6.5
Tire Grade Natural Rubber Quality Overview and
Suggested Equivalencies

Quality (Subjective)	Viscosity Stabilize Grades	Technical Grades	Smoked Sheet Grades
Very high	SMRCV 50	Pale crepe	RSS1X
	TSR CV 50	TSR L	RSS1
			TSR 5
High	TSR CV 60	TSR 10	RSS2
Good		TSR20	RSS3

6.5 QUALITY

Like all raw materials in the tire factory, there are three primary indicators of natural rubber quality: i) viscosity, including uniformity and consistency, ii) no contamination, including high levels of fatty acids, and iii) packaging. Of the three quality parameters of natural rubber, viscosity is the most important, followed by fatty acids which can bloom to the surface of tire components and detrimentally impact on tack and adhesion.

Viscosity is a function of the elastomer's molecular weight, molecular weight distribution, and the concentrations of other materials present in the polymer, such as low-molecular-weight resins, fatty acids, and other natural products. Viscosity impacts the initial mixing of the rubber with other compounding ingredients and subsequent processing of the compounded materials to form the final manufactured product (1).

Natural rubber viscosity is related to two major factors: (i) viscosity of the rubber produced by the specific rubber tree clone, and (ii) the viscosity stabilization method. A variety of methods are available to characterize the viscosity of natural rubber. The most popular is Mooney viscosity (Vr), which is obtained by measuring the torque that is required to rotate a disc embedded in the rubber or a compounded sample. This procedure is defined in ASTM D1646, entitled 'Standard Test Methods for Mooney Viscosity, Stress Relaxation, and Pre-vulcanization Characteristics (Mooney Viscometer) (8). The viscosity will typically range from 45 to over 100. The information obtained from a Mooney viscometer can include:

- Viscosity (Vr), typically measured at 100°C, provides a measure of the ease with which the material can be processed. It depends on molecular weight and molecular weight distribution, molecular structures such as stereochemistry and polymer chain branching, and non-rubber constituents. Caution is always required when attempting to establish relationships between Mooney viscosity and molecular weight. It is expressed as ML1+4 or sometimes ML1+8 (i.e. Mooney Large Rotor, with 1 minute pause and 4- or 8-minute test duration).

- Stress relaxation, which can provide information on gel (tx95), is defined as the response to a cessation of sudden deformation when the rotor of the Mooney viscometer stops. The stress relaxation of rubber is a combination of both an elastic and a viscous response. A slow rate of relaxation indicates a higher elastic component in the overall response, whereas a rapid rate of relaxation indicates a higher viscous component. The rate of stress relaxation can correlate with molecular structure characteristics, such as molecular weight distribution, chain branching, and gel content. It can be used to give an indication of polydispersity, M_n/M_w (where M_n is the number average molecular weight and M_w is the weight average molecular weight). It is determined by measuring the time for a 95% (T-95) decay of the torque at the conclusion of the viscosity test.
- Delta Mooney, typically run at 100°C, is the final viscosity after 15 minutes. This provides another measure of the processing characteristics of the rubber. It will provide a measure of the ease of processing compounds initially milled before being extruded or calendared (e.g. hot feed extrusion systems).
- Mooney Peak, which is the initial peak viscosity at the start of the test, is a function of the green strength and can be a measure of factory shelf-life of the compound.
- Compounded polymer pre-vulcanization properties or scorch resistance for the compounded natural rubber, which is conducted at temperatures ranging from 120°C to 135°C.

Much work has been done to establish a relationship between Mooney viscosity and molecular weight of natural rubber, as well as the molecular weight distribution. Bonfils and coworkers measured the molecular weight and molecular weight distribution of a number of samples of rubber from a range of clones of *Hevea brasiliensis* and noted the trend shown in Table 6.6.

Although clearly not linear, there is an empirical relationship between Mooney viscosity and M_w. The relationship between intrinsic viscosity and Mooney viscosity can be improved by mastication of the test samples, thereby improving the homogeneity and achieving reported correlation coefficient approaching 0.87 (1).

TABLE 6.6

Wallace Plasticity and Molecular Weight of Natural Rubber (9)

Sample	Po	ML1+4 (MU)	M_w (g/mol)
1	32	57	746,000
2	41	78	739,000
3	54	92	799,000
4	62	104	834,000

Mastication or milling also narrowed the molecular weight distribution, which is an important factor.

The cure characteristics of natural rubber are highly variable, due to such factors as maturity of the specific trees from which the material was extracted, the method of coagulation, pH of the coagulant, preservatives used, dry rubber content, and viscosity stabilization agent used.

A standardized formulation has been developed to enable a comparative assessment to be made of different natural rubbers and has been known as the ACS1 (American Chemical Society #1). The formulation consists of natural rubber (100 phr), stearic acid (0.5 phr), zinc oxide (6.0 phr), sulfur (3.5 phr), and 2-mercapto-benzothiazole (MBT, 0.5 phr). This formulation is very sensitive to the presence of contaminants or other materials that may be present in natural rubber, such as fatty acids, amines, and amino acids, which may influence the vulcanization rate. The formulation has been documented in ASTM standard D3184 (10).

Natural rubber is susceptible to oxidation. This can affect both the processing qualities of the rubber and also the mechanical properties of the final compounded rubber. Natural antioxidants will offer protection to degradation of natural rubber and this can be measured by changes in the material's plasticity. The Wallace Plasticity test reports two measures.

- Wallace plasticity (Po), which is a measure of the compression of a sample after a load has been applied for a defined time.
- Plasticity Retention Index (PRI), which measures recovery after a sample has been compressed, heated, and subsequently cooled. PRI% is defined as $(P_{30}/Po) \times 100$, where Po is the original Wallace plasticity and P_{30} is the plasticity after aging for 30 minutes at typically 140°C. During processing in, for example, a tire factory, natural rubber with low PRI values tends to break down more rapidly than rubber with high values.

Various equations have been proposed, which provide an empirical relationship between Mooney viscosity (Vr) and Wallace Plasticity, Po. These equations depict a linear relationship between these two parameters and are, therefore, typically of the form:

$$Vr = X \, Po + \text{Constant C} \tag{6.1}$$

The numerical coefficient, X, and the constant, C, are a function of the clone and grade of rubber, respectively, but normally fall between 1.15–1.50 for the coefficient X and 4.0–12.5 for the constant C in Equation 6.1 (11).

There are additional materials, which can be added to assist in improving the processability of natural rubber. These include the use of peptizers, such as 2,2′-dibenzamidodiphenyl disulfide, which, when added at levels of around 0.25 phr, can significantly improve productivity of the mixers, allow lower mixing temperatures, improve mixing uniformity, and reduce mixing energy. Synthetic polyisoprene, when added at levels up to 50% of the total polyisoprene content, will also reduce viscosity with little loss in other compound mechanical properties. It also allows for better control of component tack, which is important in subsequent product assembly steps, such as occur in tire building.

In many respects, the end-user quality requirements for natural rubber are similar to those for synthetic elastomers, such as SBR, polybutadiene (BR), or butyl rubbers. Quality parameters of importance can therefore include:

- Consistency. Within a grade, end-consumers of natural rubber require more uniformity, less spread in properties such as plasticity retention index, and a desire to eliminate the need for warming (placing in a heated room prior to mixing) of the rubber prior to mixing.
- Uniformity in Compounding. In tire and industrial goods manufacturing, natural rubber uniformity (e.g. with respect to viscosity) is required for final compound consistency and consistent downstream processing characteristics, such as extrusion.
- Packaging. Bales must be wrapped properly to ensure no moisture penetrates and to prevent mold growth. Adhesion of bales during shipping and storage must be prevented. Material supply reliability includes correct box weights, labeling, and other appropriate identification information, such as certificates of analysis and material safety data sheets (MSDS), depending on the regulatory requirements and policies of the receiving plant location.
- Contamination. Considerable work has been done at lowering the dirt level in both technically specified and visually inspected rubbers. As noted earlier, the last revision to the Standard Malaysian Rubber, SMR, scheme introduced the following revisions:
  - Dirt levels specifications were reduced from 0.10 to 0.08 wt%, and from 0.20 to 0.16 wt% for SMR 10CV and SMR 20CV, respectively.
  - Constant viscosity (CV) grades of SMR5 were defined with viscosities of 50 and 60, each within a ± 5 range (SMR 50CV, SMR 60CV). Dirt levels of 0.03% are now typical.
  - In addition, avoidance of other foreign material, moisture, and degraded polymer is to be ensured.
- Fatty Acids. Excessive levels of fatty acids, such as palmitic acid, oleic acid, and stearic acid, can bloom to the surface of compounded rubber components prepared for tire building or other engineered product assembly. Tire manufacturing plants may have component tack difficulties when, for example, a TSR 20 with fatty acid levels of 0.30 wt % is changed to a TSR 20 grade with a fatty acid level of 0.9–1.0 wt %. This may be due to bloom. High levels of fatty acids can also influence vulcanization kinetics. Tack-inducing resins, such as ExxonMobil Escorez 1102, may also be used to mitigate high levels of bloom.

Fatty acid levels, to a large degree, are a function of the amount of washing the raw materials undergoes prior to shipping. Malaysian technical specified rubbers (SMR) are produced to clearly defined dirt levels and thus require little washing. In consequence, fatty acid levels can be relatively high. However, other regional sources, such as Indonesian rubber, may initially contain much higher dirt levels, require more washing, and, as a result, have a higher concentration of fatty acids removed before bailing and shipping. Fatty acid concentrations are also important in ensuring consistent curing.

6.6 PROPERTIES OF NATURAL RUBBER COMPOUNDS

The molecular weight of natural rubber (M_w) is of the order of 750,000–900,000. Consequently, natural rubber compounds tend to have high tensile and tear strength. Two ASTM formulas can be used to screen not only different grades of natural rubber but also for fundamental filler systems and cure system studies (Table 6.7).

Compound 1 is a gum formula containing the natural rubber grade of interest, 6.0 phr zinc oxide, 0.5 phr stearic acid, and a conventional cure system consisting of 0.5 phr mercaptobenzothiazole (MBT), and 3.5 phr sulfur. Compound 2 is a filled compound containing 35 phr of high-abrasion furnace (HAF) carbon black. For carbon black screening, the filler loading is increased to 50 phr, and the accelerator, MBT, is replaced with the disulfide, mercaptobenzothiazole disulfide (MBTS). These base formulas have become established as recognized compounds in many of the international standards (12, 13).

To illustrate the properties of Compound 1, comparative data are presented in Table 6.8 where MBT has been replaced with diphenylguanidine (DPG) and cyclohexyl-2- benzothiazolesulfenamide (CBS) at equal phr. CBS tends to show the highest state of cure (rheometer torque), tensile strength, and tear strength, and the lowest compression set.

Similar representative data can be prepared for filled compounds, using Compound 2 in Table 6.9. Three accelerators, CBS, t-butyl-2-benzothiazolesulfenamide (TBBS), and 2-(4-morpholinothio)-benzothiazole (MBS), were evaluated at 1.00 phr (Table 6.9) and basic mechanical properties, such as tensile strength, tear strength, and aged properties were measured.

CBS tends to produce superior mechanical properties, demonstrating why it is the preferred primary accelerator in natural rubber compounds. For example, using CBS, a tensile strength of 28 MPa is obtained, which compares with 24 and 23 MPa for MBS and TBBS, respectively. For this reason, CBS tends to be the preferred accelerator for natural rubber-based radial truck tire tread compounds. Truck tire tread wear, depending on the specific service, is a function of the compound tensile strength, tear strength, and resistance to thermo-oxidative degradation. High-molecular-weight natural rubbers, such as RSS2, will therefore be used with highly

TABLE 6.7
Model Compounds for Natural Rubber Grade Evaluation (12, 13)

Compound	1	2
Natural rubber	100.00	100.00
Carbon Black (N330)		35.00
Zinc oxide	6.00	5.00
Stearic acid	0.50	0.50
Sulfur	3.50	2.25
Mercaptobenzothiazole (MBT)	0.50	0.70

TABLE 6.8

Representative Data for Model Natural Rubber Gum Compound (13, 14)

Compound	3	4	5
Technically specified rubber 5 (TSR 5)	100.00	100.00	100.00
Zinc oxide	6.00	6.00	6.00
Stearic acid	0.50	0.50	0.50
Sulfur	3.50	3.50	3.50
MBT	0.50		
DPG		0.50	
CBS			0.50
ODR rheometer			
MH–ML (delta torque)	21.50	14.50	26.50
T-90	20.00	30.00	25.25
Tensile strength (MPa)	6.35	11.00	15.60
Elongation (%)	630	600	625
300% Modulus (MPa)	1.25	1.40	2.00
Tear Die B (kN m^{-1})	35	47	54.5
Hardness (Shore A)	32	33	40
Compression set (100°C, 70 hours)	87.9	90.3	70.6

reinforcing SAF or ISAF grades of carbon black, minimum levels of oil, and, with CBS, the semi-EV cure system, which unlike in EV or efficient vulcanization producing monosulfidic crosslinks, semi- EV has nominally more disulfidic crosslinking and some polysulfidic crosslinks.

Since natural rubber is an unsaturated elastomer, it is readily susceptible to oxidation. This is reflected in the loss in tensile strength with aging at 100°C for 24 hours. For compounds cured with the accelerators CBS, MBS, and TBBS, tensile strength losses of 32%, 15%, and 33%, respectively, have been noted. Addition of antioxidants and antiozonants, such as polymerized 2,2,4-trimethyl-1,2-dihydroquinoline (TMQ) and dimethylbutyl-N-phenyl-p-phenylenediamine (6PPD) can correct such loss in mechanical properties.

6.7 SYNTHETIC ELASTOMERS USED IN TIRES

As is the case in estimates of global tire production, reports on the production and consumption of synthetic rubbers used in tires tend to vary. The following assessment (Table 6.10) is therefore a compilation of multiple data sets and has produced a reasonable view on both current and future production. It is believed that the underlying variability in production estimates is due to the global excess capacity, with more coming

TABLE 6.9
CBS, MBS and TBBS in Filled Natural Rubber Compounds (13)

Compound	6	7	8
Technically specified rubber 5 (TSR 5)	100.00	100.00	100.00
Carbon black (N330)	35.00	35.00	35.00
Zinc oxide	5.00	5.00	5.00
Stearic acid	2.00	2.00	2.00
Sulfur	2.25	2.25	2.25
CBS	1.00		
MBS		1.00	
TBBS			1.00
ODR rheometer			
MH−ML	75.0	73.0	77.5
T90	4.85	5.85	5.35
Cure rate index	42.1	31.74	42.53
Tensile strength (MPa)	28.57	24.46	22.76
Elongation (%)	467	435	392
300% Modulus (MPa)	13.65	13.25	14.75
Tear strength (KNm^{-1})	93.5	103.2	98.3
Hardness (Shore A)	67	64	62
Aged 100°C, 24 hours			
Tensile strength (MPa)	19.4	20.8	15.12
Elongation (%)	336	350	235
200% Modulus (MPa)	9.6	8.85	11.77

on-line in the next several years. The elastomers used in tires include SBR, polybutadiene, and butyl rubbers. In addition, small amounts of EPDM are used in applications such as white sidewall compounds and in blends with butyl for tire inner-tubes.

TABLE 6.10
Global Annual Total Rubber Production (1000 Tons) Estimated Through to 2027 (15) (Combined natural and synthetic rubber)

2007	22,598
2012	24,766
2017	28,483
2022	32,935
2027	37,090

TABLE 6.11

Synthetic Rubber Production (15, 16)

Year	Synthetic Rubber Production (millions of tons)
2014	14.18
2015	14.51
2016	14.84
2017	15.15
2018	15.43

Polychloroprene, nitriles, and specialty polymers, such as the fluoro-elastomers have not found any niche in the field of tire material science and technology. Global synthetic rubber consumption is expected to grow substantially over the next ten years, with that assumption most likely being the primary reason for new capacity investments. Current synthetic rubber production is of the order of 15.5 million tons (Table 6.11), which is well below the global capacity of 20.5 million tons (2, 15).

Of the total global synthetic rubber capacity, 56% is located in Asia, 16% in North America, and 14% in the European Union region (Figure 6.5). Given this, production capacity can be further considered in terms of blocks, with China and the United States being the largest producers, followed by Japan, South Korea, and Russia. The third tier would then include countries in the European Union and Taiwan.

Classification of synthetic rubber has historically been governed by the International Institute of Synthetic Rubber Producers (IISRP) and associated institutes, such as the American Society for Testing Materials (ASTM). In the case of SBR, polyisoprene rubber, and polybutadiene (BR), a series of numbers have been assigned which classify the general properties of the polymer (17, 18). For example, the IISRP 1500 series defines cold emulsion-polymerized (i.e., below 10°C), non-pigmented SBR. The 1700 series of polymers describes oil-extended cold emulsion

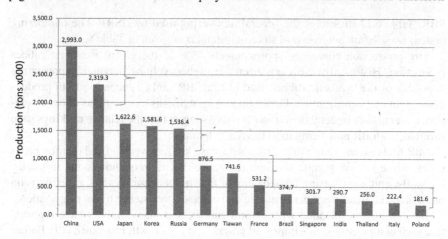

FIGURE 6.5 Global Synthetic Rubber Production (15, 16). All countries increased production versus 2016 but with India showing the only Change in ranking, replacing Thailand. Production can be split into 4 segments.

TABLE 6.12

Classification Summary of Emulsion Elastomers by the International Institute of Synthetic Rubber

Series	Description
1000	Hot non-pigmented emulsion SBR Polymerized above 38°C
1500	Cold non-pigmented emulsion SBR Polymerized below 10°C
1600	Cold non-pigmented emulsion SBR Carbon black/oil masterbatch/14 phr oil max.
1700	Oil extended cold emulsion SBR
1800	Cold non-pigmented emulsion SBR Carbon black/Oil masterbatch. More than 14 phr oil
1900	Emulsion resin rubber master batches

TABLE 6.13

Classification Summary of Solution Elastomers by the International Institute of Synthetic Rubber Producers

Category	Butadiene & Copolymers	Isoprene & Copolymers	Butadiene, Isoprene & Styrene Copolymers
Dry polymer	1200–1249	2200–2249	2250–2599
Oil extended	1250–1299	2250–2299	
Black masterbatch	1300–1349	2300–2349	
Oil-black masterbatch	1350–1399	2350–2399	
Latex	1400–1449	2400–2449	
Miscellaneous	1450–1499	2450-2499	

SBR. Table 6.12 illustrates the general numbering used by IISRP. The numbering system for solution polymerized stereo-elastomers is given in Table 6.13.

Tire production consumes approximately 75% of the global synthetic rubber production. Historically, SBR has been the highest-volume polymer, representing over 65% of the synthetic rubber used in tires. BR ranked second in both production output and consumption. Depending on how SBR is classified, as either a solution, an emulsion, or combined, reports today suggest that the relative rankings have inverted, with BR now being used in greater tonnage.

SBR finds extensive use in tire treads because it offers wet skid and traction properties, while retaining good abrasion resistance. BR is frequently found in treads, sidewalls, and some casing components of the tire because it offers good abrasion resistance and tread wear performance and enhances resistance to cut propagation.

BR can also be blended with natural rubber, and many authors have reported that such compositions give improved fatigue and cut growth resistance (19). Before reviewing elastomer characteristics required to meet any given set of tire performance parameters, it is appropriate to identify two means by which a materials engineer may describe a polymer: polymer macrostructure and polymer microstructure. The effect of polymer macrostructure and microstructure on tire compound properties and tire performance has been described in considerable detail by Colvin (18) and Rodgers et al. (20), to which further reference is recommended.

The BR market size is expected to grow from USD 10.8 billion in 2019 to USD 13.8 billion by 2024, at a CAGR of 5.1% during the forecast period (15), driven by tire, polymer modification, and industrial rubber manufacturing industries. BR can be produced from a variety of feedstocks, such as from crude C4 extractions, ethylene dimerization, butadiene, and dehydrogenation of butanes (Figure 6.6).

In all cases leading to the production of butadiene, the polymerization process entails addition of catalysts such as *n*-butyl lithium to the pure monomer, polymerization, termination *via* addition of a "short-stop" agent and an antioxidant, passing the polymer cement (polymer in hexane solution) to a flash drum and steam stripper, allowing hexane recycling, and then on to final finish and baling (Figure 6.7). Depending on the type of BR, defined by its microstructure, different catalysts will be used based on lithium for medium *cis*-butadiene levels, or titanium, cobalt, neodymium, and nickel for high *cis*-butadiene content.

The microstructure vinyl-butadiene content monomers in the polymer is controlled by the addition of a polar modifier such as tetramethylethylenediamine or TMTDA. On completion of the polymerization, finishing of the polymer in typical 33 kg bales is where most effort is required to ensure quality (Figure 6.8).

Polymers are produced by one of two chemical processes, addition or condensation polymerization. Addition polymerization occurs by one of three mechanisms, namely radical (e.g., low-density branched polyethylene), cationic (e.g., butyl rubbers), or anionic (e.g., polystyrene). Condensation polymerization of adipic acid and hexamethylene diamine, with the elimination of water, is used to produce

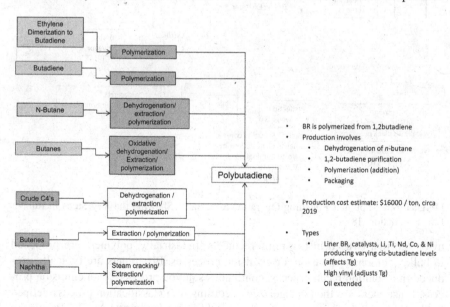

FIGURE 6.6   Butadiene Feedstock Sources

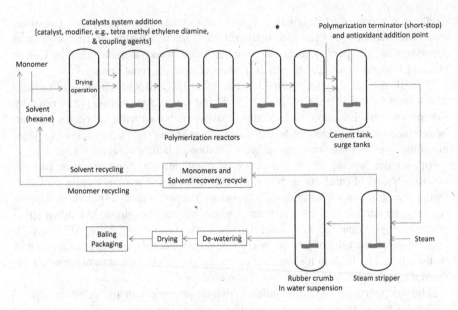

FIGURE 6.7   Solution Polymerization of Polybutadiene and Styrene Butadiene Copolymers (18)

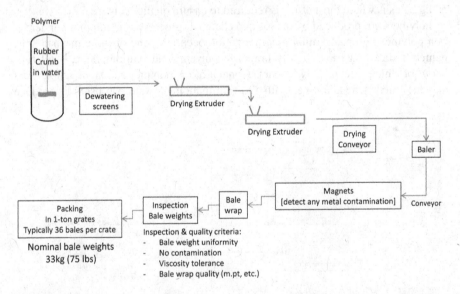

FIGURE 6.8   Polymer Finishing Operation Schematic, Showing the Key Steps in Rubber Bale Production (18)

nylon 6,6 that is used in tire reinforcements. Industrially, polymers are produced in bulk, solution, suspension, or emulsion processes. Elastomers are typically produced in a solution batch process, continuous solution process, or an emulsion process. Elastomers are then categorized according to a classification process defined by the International Institute of Synthetic Rubber Producers and ASTM. Briefly, an emulsion polymerized SBR polymer, such as IISRP SBR1500, has a bound styrene level of 23.5%, and bale Mooney viscosity of 50–55 Mooney units. The oil-extended version is classed as SBR 1712. Polymer/oil/carbon black masterbatches fall under either 1600 series or 1800 series classes (Table 6.12). Solution polymers similarly are classed as 1200 series for SBR and BR, and 2200 series for polyisoprene and copolymers of isoprene (Table 6.13).

Tires are the largest consumer of synthetic rubber. Automotive components and tires together account for nearly 70% of synthetic rubber consumption. Additional consumption is found in miscellaneous mechanical goods, plastic composites, and construction applications, such as roofing, wire and cable covers, and adhesives. For SBR specifically, passenger tire production consumes approximately 50%, with truck tires and tire retreading making up a further 20%, and the balance being in specialty tires, automotive, and non-automotive components. BR consumption is similar to SBR, with tires accounting for nearly 75% of total polymer production.

A series of empirical guidelines have been developed which might be used in designing a polymer or blend of polymers to meet a set of tire performance targets (19). By preparing a series of blends of BR and SBR, a range of polymer Tg values can be obtained which vary from as low as −100°C to −20°C. Evaluation of these blends has allowed the deduction of the following observations:

1. As Tg increases, there is a near-linear decrease in abrasion resistance.
2. Wet grip or traction improves, again near-linearly, with an increase in compound Tg.
3. Addition of styrene leads to an increase in traction performance and a loss in abrasion resistance.

When the dynamic stiffness or storage modulus of a tread compound is measured from −100°C to +100°C, the resulting tangent delta or $tan\ \delta$ curve can be divided into segments or zones, which will characterize the tread compound performance in a tire (Figure 6.9). The tan delta value in the region of the compound glass transition temperature (Tg) will define its abrasion characteristics. Ice, snow, and wet traction properties can be predicted by the tan delta values in the region of −20°C to 0°C, while rolling resistance is determined by tan delta values in the region of +60°C.

The tan delta of a compound is a function of the storage modulus measured in either strain (E') or shear (G') and the hysteresis or loss modulus (E'' in strain and G'' in shear)

$$\text{Tangent delta } (\tan \delta) = E'' / E' \text{ (or } G'' / G') \qquad (6.2)$$

Plotting this relationship with temperature produces the characteristic graph in Figure 6.9. Qualitatively, as the temperature increases above the compound Tg, $tan\ \delta$ will decrease because, as the compound gets warmer, at the molecular level there is greater polymer chain movement, and the compound softens, leading to an increase in energy loss (E'') and a decrease in stiffness (E'). Conversely, as the temperature drops below the Tg, the molecular chains are frozen, there is less molecular movement, and stiffness increases, leading to a rapid decrease in $tan\ \delta$ (Figure 6.10).

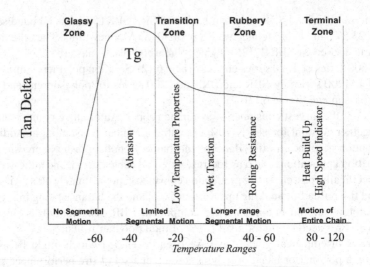

FIGURE 6.9 Tangent Delta Temperature Profile for Predictive Tire Tread Compound Performance

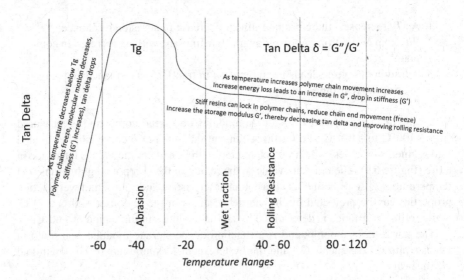

FIGURE 6.10 Compound Dynamics

High-*Tg* resins in a compound formulation, if near-miscible at the molecular level, can control the loss in storage modulus at temperatures above 30°C, thereby adjusting *tan δ*, and potentially allowing rolling resistance improvements without the typical trade-off in traction performance. Though the fundamental understanding of the *tan δ* curve is complex, a qualitative understanding such as this can allow a better-defined development strategy when developing or tuning a compound formulation for a specific tire design.

TABLE 6.14

New Generation Solution SBR

Generation	Descriptions and Definitions	Function
1	Linear copolymer of styrene and butadiene	Tg adjusted with styrene and vinyl butadiene levels (TMEDA addition)
2	Coupled copolymers. Tin or silicon tetrachloride	Base Mooney ML1+4 of 20 jumping to 80 on coupling
3	Epoxy groups to improve polymer filler interaction	Carbon black interaction
4	Mono-terminal (only one chain end) alkoxysilane functionalization	Silica interaction
5	Bi-functional (both chain ends) alkoxysilane groups	Reduce Payne effect, improve polymer -filler interaction

To improve polymer performance, addition of functional groups has been the focus of considerable research. To this end, a new series of BR and solution SBR rubbers has emerged, where functionalization at the chain end has allowed better polymer filler interactions, with corresponding improvement in compound properties, specifically abrasion resistance and hysteresis. In the case of solution SBR, the development has been described in terms of "generations", 1 through 5 or sometimes referred to as 0 through 4 which emerged only recently. Table 6.14 describes the new generations of solution SBRs that have been commercialized.

In summary, the core quality requirement remains unchanged for later downstream tire manufacturing operations, and, as is the case for natural rubber, centers on the following factors:

1. bale weight consistency
2. bale wrap melt temperature, dispersion, seal gauge, and absence of any tearing which can cause bales to stick together
3. Mooney viscosity specification and tolerance
4. Absence of gel
5. No discoloration
6. No contamination, no moisture.

6.8 CARBON BLACK AND ITS PROPERTIES

Fillers, or reinforcement aids, such as carbon black, clays, and silicas, are added to rubber formulations to meet material property targets, such as tensile strength. Considering first carbon black, it can be described qualitatively by a series of properties: particle size (and surface area); particle size distribution; structure (particle aggregates); and surface activity (chemical functional groups such as carboxyl and ketones). Key properties describing a carbon black can be listed as described in

(Table 6.15. The measurable properties, such as iodine number and dibutyl phthalate (DBP) absorption can be further tabulated, giving precise descriptions for each grade of carbon black(Table 6.16).

Further reference can also be made to ASTM Standards D1566-04 for definitions of general compounding terms (21) and to D3053-04, specifically for carbon blacks (22). As an empirical guide, an increase in a carbon black aggregate size or structure will result in an improvement in cut growth and fatigue resistance. A decrease in particle size results in an increase in abrasion resistance and tear strength, a decrease in resilience, and an increase in hysteresis and heat build-up. In addition, there are a number of criteria relating carbon black to the hysteretic properties of rubber compounds (19). These included loading, aggregate size, surface area, aggregate size distribution, aggregate irregularity (structure), surface activity, dispersion, and phase distribution within a heterogeneous polymer system. Briefly:

1. Reduction of carbon black loading lowers tire rolling resistance. At a constant carbon black concentration, an increase in oil level will increase rolling resistance but also improve traction (at low oil levels, an increase in oil level may decrease compound hysteresis by improving carbon black dispersion).
2. Increasing carbon black fineness used in a compound increases the contribution to both rolling resistance and traction.
3. An increase in the broad aggregate size distribution decreases the tire rolling resistance with constant surface area and DBP.
4. Tread-grade carbon blacks can be selected to meet defined performance parameters of damage resistance and tear strength, rolling resistance, traction, and wear

As carbon black level increases, there are increases in compound heat build-up and hardness, and, in tires, an increase in rolling resistance and wet skid properties. Tensile strength, compound processability, and abrasion resistance, however, pass through an optimum, after which these properties deteriorate.

In tire manufacturing plants, there are typically around 6–8 grades of carbon black, this restriction being due to the installation costs of storage towers or silos, pneumatic conveying systems, and injection scales at the Banburies. Grades could include two tread compound grades (N121, N234), a sidewall grade (N330), a wire coat or skim compound grade (N326, N347, or N551), shoulder wedge and apex grade (N550), and an inner liner compound grade, almost universally N660.

6.9 SILICA, SILICATES, AND OTHER INORGANIC FILLERS

Addition of silica to a rubber compound offers a number of advantages, such as improvement in tear strength, reduction in heat build-up, and increase in compound adhesion in multicomponent products, such as tires. The manufacturing process has been illustrated in Figure 6.11. Two fundamental properties of silica and silicates influence their use in rubber compounds, namely ultimate particle size and the extent

TABLE 6.15

Carbon Black Terms

Furnace Carbon Black	Class of carbon blacks produced by injection of defined grades of petroleum feedstock into a high-velocity stream of combustion gases under a set of defined processing conditions, e.g., N 110 to N762.
Thermal Carbon Black	Type of carbon black produced by thermal decomposition of hydrocarbon gases, e.g., N 990, N 991.
Acetylene Carbon Black	Type of high-structure carbon black. Can improve thermal and electrical conductivity.
Microstructure	Describes the arrangement of carbon atoms within a carbon black particle.
Particle	Small spherical component of a carbon black aggregate produced by fracturing the aggregate. Particle size is measured by electron microscopy.
Aggregate	Distinct, colloidal mass of particles in its smallest dispersible unit.
Agglomerate	Arrangement or cluster of aggregates.
Structure	Measure of the deviation of the carbon black aggregate from a spherical form.
Iodine number	Weighting of iodine absorbed per kg of carbon black. Measure of particle surface area. The smaller the particle size, the greater the iodine number.
Carbon black DBP	Volume of dibutyl phthalate (cm^3) absorbed by 100 g of carbon black. DBP number is a measure of the structure of the carbon black aggregate.
Tint	Tint is a ratio of the reflectance of a reference paste to that of a sample paste consisting of a mixture of zinc oxide, plasticizer, and carbon black.
CTAB	Measure of the specific surface area corrected for the effect of microspores. CTAB (cetyltrimethyl ammonium bromide) is excluded from the smaller interstices and thus better represents the portion of a particle surface area in contact with the polymer.
Nitrogen surface area	Measure of total particle surface area, possibly due to nitrogen gas being able to cover the full surface, including pores without interface from surface organic functional groups.
Compressed DBP	The DBP test, but where the sample undergoes a series of compressions (4 times to 24,000 lb) before testing. This enables a measure of changes the carbon black will undergo during compound processing.
Pellet	Mass of compressed carbon black formed to reduce dust levels, for ease of handling, and improved flow.
Fines	Quantity of dust present in a pelletized carbon black; should be at the minimum level possible.
Pellet hardness	Measure of the load in g to crush a defined number of pellets.
Ash	Residue remaining after burning carbon black at 550°C for 16 hours.
Toluene discoloration	Hydrocarbons extractable in toluene from carbon black; can be used as a measure of the residence time in a furnace.
Hydrogen and oxygen	Residual hydrogen and oxygen remaining after carbon black is produced. It can affect compound vulcanization kinetics and reinforcement potential.

of hydration. Other physical properties, such as pH, chemical composition, and oil absorption, are of secondary importance.

Silicas, when compared to carbon blacks of the same particle size, do not provide the same level of reinforcement, though the deficiency of silica largely disappears when coupling agents are used with silica. Unlike carbon black, there is

TABLE 6.16
Properties of Carbon Black

ASTM Designation	Iodine Number	Oil Absorption Number	Compressed Oil Absorption Number	NSA Multipoint	STSA	Tint Strength
N110	145	113	98	126	115	124
N115	160	113	96	137	124	123
N120	120	114	99	122	114	129
N121	121	132	112	121	114	119
N125	117	104	89	122	121	125
N134	142	127	103	143	137	131
N220	121	114	100	119	106	115
N231	121	92	86	108	107	117
N234	120	125	100	119	112	124
N299	108	125	105	104	97	113
N326	82	72	69	78	76	112
N330	82	102	88	82	75	103
N339	90	120	101	93	88	110
N347	90	124	100	85	83	103
N351	68	120	97	73	70	100
N358	84	150	108	80	87	98
N375	90	114	97	96	78	115
N550	43	121	88	42	39	
N630	36	78	62	32	32	
N650	36	122	84	36	35	
N660	36	90	75	35	34	
N762	27	65	57	29	28	
N772	30	65	58	32	30	
N990		38	37	9	8	
N991		35	37	8	8	

no standardized set of descriptors for silicas, such as seen in Tables 6.15 and 6.16. Addition of silica to a tread compound leads to a loss in tread wear, even though improvements in hysteresis and tear strength are obtained. The tread wear loss can be corrected by the use of silane coupling agents. The chemistry of silica can thus be characterized as follows:

1. Silica, which is amorphous, consists of silicon and oxygen arranged in a tetrahedral structure of a three-dimensional lattice. Particle size ranges from 1 to 30 nm and surface area from 20 to 300 m^2/g. There is no long-range crystal order, only short-range ordered domains in a random arrangement with neighboring domains.
2. Surface silanol concentration (silanol groups, -Si-O-H) influence the degree of surface hydration.

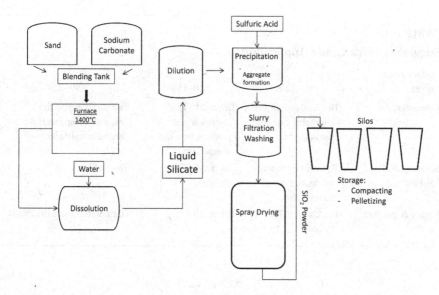

FIGURE 6.11   Simplified Schematic Illustrating Silica Production Processes

3. Silanol types fall into three categories – isolated, geminal (two -OH hydroxyl groups on the same silicon atom), and vicinal (two -OH hydroxyl groups on adjacent silicon atoms) – as illustrated in Figure 6.12.
4. Surface acidity is controlled by the hydroxyl groups on the surface of the silica and is intermediate between those of P-OH and B-OH. This intrinsic acidity can influence peroxide vulcanization, although in sulfur curing, there is no significant effect. Rubber-filler interaction is affected by these sites.
5. Surface hydration, caused by water vapor absorption, is affected by surface silanol concentration. High levels of hydration can adversely affect final compound physical properties. Silicas are hygroscopic and thus require dry storage conditions.

When tread compound silica loading approaches 20%, the decrease in abrasion resistance of a compound renders the formulation no longer practical. Silane coupling agents offer the potential to overcome such decreases in compound performance. Silicas can be divided into groups or classes, i.e., standard or conventional silicas, semi-highly dispersible (semi-HD) or easily dispersible silica, with the latest group developed being termed highly dispersible silica or HDS (Table 6.17). The silanol composition on the surface of the first three types of silicas remains to be elucidated, but it would be anticipated that the HDS silicas would have higher concentrations of geminal groups, whereas the conventional silica would have a greater concentration of isolated silanols (19) (Figure 6.12).

To achieve the required performance of silica in a tread or other tire compound, a silane coupling agent is needed. There are three silane coupling agents of commercial significance and these have similar properties: mercaptopropyltrimethoxysilane

TABLE 6.17

Suggested Silica Grade Applications

Surface Area (m²/g)	90–130	130–180	180–220
Conventional	Tire casing. Non-tire internal & external components	Tire treads. Tire casings & non-tire products external components	Tire treads & non-tire external components for abrasion resistance
Semi highly dispersible	Tire casing. Non-tire products, external components	Tire treads. Tire casings	Tire treads
Highly dispersible	Tire casings. Tire treads	Tire treads	High-performance tire treads

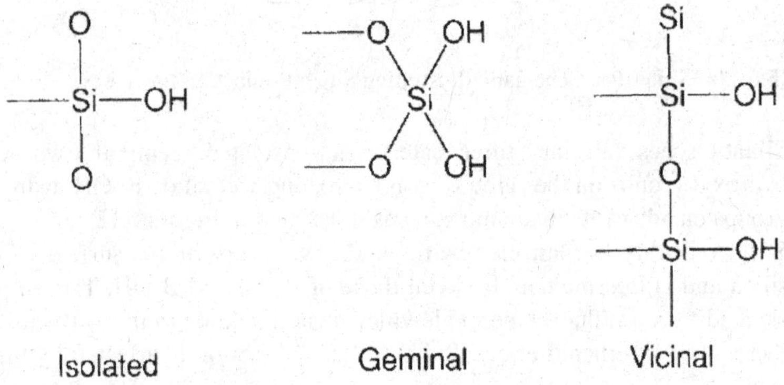

Isolated   Geminal   Vicinal

FIGURE 6.12   Silanol Chemistry

(A189), bis(triethoxysilyl)ethyltoluene) polysulfide (Y9194), and bis(3-triethoxysilyl-propyl) tetrasulfane (TESPT), with commercial designations being in parentheses. The coupling agent TESPT is the preferred silane coupling agent and has been covered more extensively in the literature than the other silane coupling agents; however, the following discussion on the use of silane coupling agents is applicable to all three materials. TESPT, a bifunctional polysulfidic organosilane, was introduced as a coupling agent to improve the reinforcement properties of silicas in rubbers. Silane coupling agents perform two functions: i) improve dispersion measured by the drop in the Payne Effect, and ii) create crosslinks between the silica surface and the polymer, i.e., improve the polymer–filler interaction. Use of coupling agents thus offers the following advantages:

1. Lowers heat build-up and hysteresis in silica-loaded compounds
2. Increases 300% modulus and tensile strength, again, in silica-loaded compounds
3. Improves reinforcing effect of clays and whiting, in addition to that for silica

4. Serves as a reversion resistor in equilibrium cure systems
5. Improves abrasion resistance

The mechanism of silane coupling agent reinforcement comprises two phases, namely i) the silanization reaction, in which coupling agent reacts with silica, and ii) the formation of crosslinks between the modified silica and the polymer. Silanization of the silica surface can occur quite readily, though, with TESPT systems, the reaction is generally carried out *in situ* at between 150 and 160°C in an internal mixer. Though an excess of silanol groups are present on the silica surface and reaction rates are fast, this high temperature is required because of the steric hindrance around the silylpropyl group in TESPT. Essentially, three types of functional silanol groups exist on the silica surface: isolated hydroxyl groups, geminal groups (two -OH groups on one Si atom), and vicinal groups. The silanization reaction is illustrated in Figure 6.13.

The filler/silane intermediate can now react with the allyl position of unsaturated sites on the polymer chain. The vulcanization of rubber is known to proceed *via* the reaction with an accelerator, such as a sulfenamide, with sulfur, zinc oxide, or stearic acid, to generate a sulfurating agent. On completion of the reaction, the pendant accelerator will cleave off (i.e., Captax) after generation of a crosslink. This accelerator residue, Captax, is an accelerator in its own right, and continues to participate in further crosslinking as vulcanization continues.

In silica reinforcement systems containing TESPT, the reaction is similar when the TESPT/silica intermediate is present instead of sulfur, in which case the crosslinking agent is the polysulfidic sulfur chain. Wolff showed that mercaptobenzothiazyl disulfide (MBTS) reacts with the tetrasulfane group, thus forming 2 moles of the polysulfide (33)

The silica particle is on one side and the -mercaptobenzothiazole on the other. This polysulfidic pendant group on the silica surface will now undergo crosslink formation with the polymer, in much the same way as occurs in rubber-bound intermediates that convert to crosslinks. Wolff (33) suggested that the MBT entity reacts with the allyl position of a double bond of the rubber, thus releasing MBT and forming the rubber–silica bond (Table 6.17).

A series of additional filler systems include kaolin clay (hydrous aluminum silicate), mica (potassium aluminum silicate), talc (magnesium silicate), limestone (calcium carbonate), and titanium dioxide. As with silica, the properties of clay can be enhanced through treatment of the surface with silane coupling agents. Thioalkylsilanes can react with the surface to produce a pendant thiol group, which may react with the polymer through either hydrogen bonding, van der Waals forces, or crosslinking with other reactive groups. Titanium dioxide finds extensive use in white products such as white tire sidewalls, where appearance is important.

6.10 PROTECTANT SYSTEMS

The presence of carbon–carbon double bonds renders elastomers susceptible to attack by oxygen and ozone, and also to thermal degradation. An excellent review of the chemistry and compounding of rubber formulations for best oxidation and

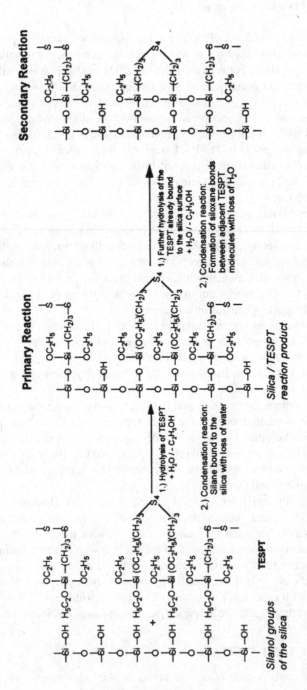

FIGURE 6.13 Reaction of TESPT with Silica and Polymer

degradation resistance has been published by Sung Hong, to which further reference is recommended (11)

Oxidation of elastomers is accelerated by a number of factors, including heat, heavy metal contamination, sulfur, light, moisture, swelling in oil and solvents, dynamic fatigue, oxygen, and ozone. Three variables in a tire compound formulation can be optimized to resist degradation: use of waxes for short term protection, antioxidants, and antiozonants.

Para-phenylenediamine antidegradants are used to hinder the oxidative activity of metal ions. A major cause of failure in rubber products is surface crack development. The growth of such cracks under cyclic deformation results in fatigue failure. Fatigue-related cracks are initiated at high stress zones. Attack by ozone can induce crack initiation at the surface which then propagates as a result of flexing. Ozone-initiated cracking can be seen as crazing on the sidewalls of old tires. Ozone readily reacts with the carbon–carbon double bonds of unsaturated elastomers to form ozonides. Under strain, ozonides readily decompose, resulting in chain cleavage and a reduction in polymer molecular weight. Such polymer molecular weight reduction becomes apparent as surface crazing and cracking (Figure 6.14).

The selection criteria governing the use of antidegradants can be summarized as follows:

1. Discoloration and staining: For black-colored compounds in tires, this is not a significant factor, though brown staining due to phenylenediamines such as 6PPD is a focus of much research.
2. Volatility: As a rule, the higher the molecular weight of the antioxidant, the less volatile it will be, an exception being hindered phenols, which tend to be highly volatile compared with amines of equivalent molecular weight.
3. Solubility: Low solubility of an antidegradant will cause the material to bloom to the surface, with consequent loss of protection of the product.
4. Chemical stability: Antidegradant stability against heat, light, oxygen, and solvents is required for durability.
5. Concentration: Most antidegradants have an optimum concentration for maximum effectiveness, after which the material solubility becomes a limiting factor. *Para*-phenylenediamines offer good oxidation resistance at a

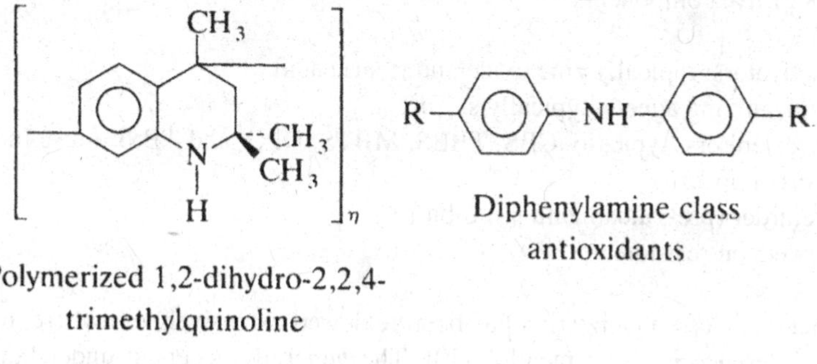

Polymerized 1,2-dihydro-2,2,4-trimethylquinoline

Diphenylamine class antioxidants

FIGURE 6.14 Two Main Classes of Stabilizers

loading of 0.5–1.0 phr and good antiozonant protection in the range 2.0–5.0 phr. Above 5.0 phr, *p*-phenylenediamines such as 6PPD may tend to bloom or migrate to the surface
6. Environment, health, and safety: For ease of handling and avoidance of dust and inhalation, antidegradants should be dust free, although free flowing.

In tire compounds, the two most common stabilizers are TMQ (antioxidant) at between 1.0 and 2.0 phr and 6PPD (antiozonants) found from between 1.5 and 4.5 phr.

Waxes are an additional class of materials used to improve rubber ozone protection, primarily under static conditions. Waxes used in elastomeric formulations fall into two categories. Microcrystalline wax has a melting point in the region 55–100°C and is extracted from residual heavy lube stock of refined petroleum. Paraffin wax has melting points in the range 35–75°C and is obtained from the light lube distillate of crude oil.

1. Short-term static protection is achieved by use of paraffinic waxes.
2. Microcrystalline waxes provide long-term ozone protection while the finished product is in storage.
3. A critical level of wax bloom is required to form a protective film for static ozone protection.
4. Optimized blends of waxes and PPDs provide long-term product protection under both static and dynamic applications and over a range of temperatures.
5. Excess levels of wax bloom can have a detrimental effect on fatigue resistance, because the thick layer of wax can crack under strain and the crack can propagate into the product.

6.11   VULCANIZATION SYSTEMS

Vulcanization, named after Vulcan, the Roman God of Fire, describes the process by which physically soft, compounded rubber materials are converted into high-quality engineering products. The vulcanization system constitutes the fourth component in an elastomeric formulation and functions by inserting crosslinks between adjacent polymer chains in the compound. A typical vulcanization system in a tire compound consists of five components

- activators, typically zinc oxide and stearic acid
- vulcanizing agents, typically sulfur;
- accelerators, typically CBS, TBBS, MBTS, DPG, and TBSI in use in tires (Figure 6.15)
- retarder (pre-vulcanization inhibitor)
- reversion resistors

The chemistry of vulcanization has been reviewed extensively elsewhere, to which further reference is recommended (10). The generally accepted understanding of vulcanization is that it occurs first by zinc oxide and stearic acid reacting to form

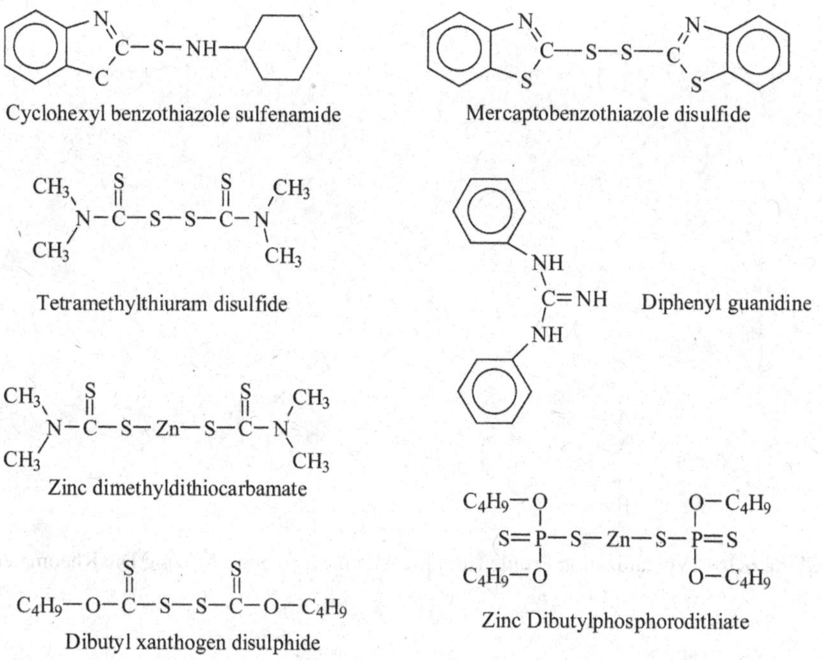

Cyclohexyl benzothiazole sulfenamide

Mercaptobenzothiazole disulfide

Tetramethylthiuram disulfide

Diphenyl guanidine

Zinc dimethyldithiocarbamate

Zinc Dibutylphosphorodithiate

Dibutyl xanthogen disulphide

FIGURE 6.15 Examples of Accelerators Used in Tire Compounds

zinc stearate. This can then readily react with an accelerator to form an activated complex, with which sulfur will react to form a sulfurating agent. This agent will then interact with the carbon–carbon double bond in a polymer chain, and then subsequently form a sulfur bridge or crosslink with another adjacent carbon–carbon double bond. The kinetics of vulcanization are typically measured using a moving-die rheometer or MRD, which produces a torque-plateau, from which vulcanization cure times can be calculated (Figures 6.16 and 6.17).

In tire material engineering, there are five factors influencing the development of cure system technology:

1. Elimination of accelerators that may generate nitrosamines
2. Accelerators for improved scorch resistance and improved ease of compound processing during final product assembly, e.g. a tire
3. Vulcanization systems suitable for high-temperature curing with associated benefits in terms of productivity
4. Improved reversion resistance
5. Improvement in compound attributes, such as mechanical or dynamic properties

Scorch safety is obtained by use of a retarder such as N-cyclohexylthiophthalimide (CTP), which is, by far, the largest retarder (in terms of tonnage) used in the rubber

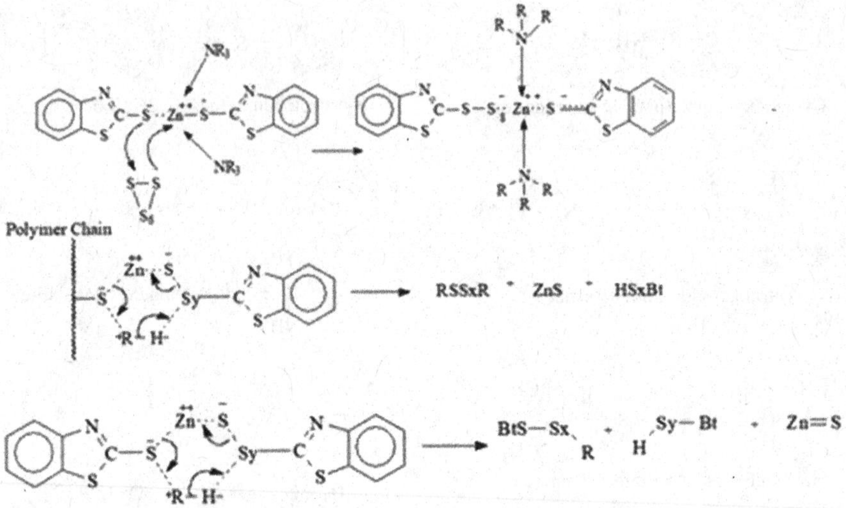

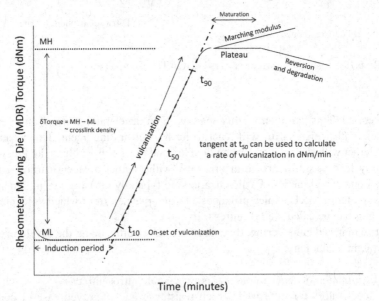

FIGURE 6.16 Vulcanization Profile from Measurements using a Moving Die Rheometer

FIGURE 6.17 Mechanism of Vulcanization

industry. It is typically added to the formulation at between 0.10 and 0.25 phr and functions by extending the vulcanization induction period (Figure 6.15).

6.12 TIRE FACTORY COMPOUND PROCESSING AIDS

Processing aids in tire compounds fall into one of three material classes: process oils which serve primarily to adjust compound viscosity, plasticizers and dispersion aids, and resins.

Oils fall into one of three primary categories: paraffinic, naphthanic, and aromatic, or, more recently, aromatic oils have been replaced by distilled aromatic extracts (TDAE). The proper selection of oils for inclusion in a formulation is important. If the oil is incompatible with the polymer, the oil will migrate out of the compound with consequent loss in required physical properties, loss in rubber component surface properties, and deterioration in component-to-component adhesion, as in a tire. The compatibility of an oil with a polymer system is a function of the properties of the oil, such as viscosity, molecular weight, and molecular composition. Aniline point is a measure of the aromaticity of an oil. It is the point at which the oil becomes miscible in aniline. Thus, the lower the aniline point, the higher the aromatic content. All three classes of oils contain high levels of cyclic carbon structures; the differences are in the number of saturated and unsaturated rings. Oils can therefore be described and compared qualitatively.

Though processing oils, waxes, and fatty acids can be considered as plasticizers, within the rubber industry, the term "plasticizer" is used more frequently to describe the class of materials which includes esters, pine tars, and low molecular-weight polyethylene. Phthalates are the most frequently used esters. Dibutylphthalate (DBP) tends to give soft compounds with tack; dioctylphthalate (DOP) is less volatile and tends to produce harder compounds because of its higher molecular weight. Polymeric esters, such as polypropylene adipate (PPA), are used when low volatility is required along with good heat resistance. Though total consumption is tending to fall, pine tars are highly compatible with natural rubber, give good filler dispersion, and can enhance compound properties such as fatigue resistance and component-to-component adhesion, which is important in tire durability. Other low-volume plasticizers include factice (sulfur-vulcanized vegetable oil); fatty acid salts such as zinc stearate, which can also act as a peptizer; rosin; low-molecular-weight polypropylene; and organosilanes, such as dimethylpolysiloxane.

Peptizers serve as either oxidation catalysts or radical acceptors, which essentially remove free radicals formed during the initial mixing of the elastomer. This prevents polymer recombination, allowing a consequent fall in polymer molecular weight, and thus a reduction in compound viscosity. This polymer softening then enables incorporation of the range of compounding materials included in the formulation. Examples of peptizers are pentachlorothio-phenol, phenylhydrazine, certain diphenylsulfides, and xylyl mercaptan. Each peptizer has an optimum loading in a compound for greatest efficiency. Peptizers such as pentachlorothiophenol are generally used at levels between 0.1 and 0.25 phr. This enables significant improvement in compound processability, reduction in energy consumption during mixing, and improvement in compound uniformity. High levels can, however, adversely affect the compound properties, as excess peptizer continues to catalyze polymer breakdown when the product is in service.

Resins fall into one of four functional categories: i) extending or processing resins, ii) tackifying resins, iii) curing resins, and iv) compound dynamic property modifiers, functioning by adjusting the compound Tg and decreasing the Payne effect. Resins have been classified in an almost arbitrary manner into hydrocarbons, petroleum resins, and phenolic resins. Hydrocarbon resins tend to have high glass transition temperatures so that, at processing temperatures, they melt, thereby allowing improvement in compound viscosity mold flow. They will, however, harden at room temperature, thus maintaining compound hardness and modulus. Within the range of hydrocarbon resins, aromatic resins serve as reinforcing agents, aliphatic resins improve tack, and intermediate resins provide both characteristics. Coumarone-indene resin systems are examples of such systems. These resins provide:

1. Improved tensile, as a result of stiffening at room temperature
2. Increased fatigue resistance, as a result of improved dispersion of the fillers and wetting of the filler surface
3. Retardation of cut growth by dissipation of stress at the crack tip (as a result of a decrease in compound viscosity).

Petroleum resins are a by-product of oil refining. Like hydrocarbon resins, a range of grades are produced. Aliphatic resins, which contain oligomers of isoprene, tend to be used as tackifiers, whereas aromatic resins, which also contain high levels of dicyclopentadiene, tend be classed more as reinforcing systems. Phenolic resins are of two types, reactive and nonreactive. Nonreactive resins tend to be oligomers of alkyl-phenyl formaldehyde, where the *para*-alkyl group ranges from C4 to C9. Such resins tend to be used as tackifying resins. Reactive resins contain free methylol groups. In the presence of methylene donors, such as hexamethylenetetramine, crosslink networks will be created, enabling the reactive resin to serve as a reinforcing resin and adhesion promoter.

6.13   COMPOUND PRODUCTION EFFICIENCIES

In a modern tire or general product production facility, rubber compounds are prepared in internal mixers. Internal mixers consist of a chamber to which the compounding ingredients are added. In the chamber are two rotors that generate high shear forces, dispersing the fillers and other raw materials in the polymer. The generation of these shear forces results in the production of a uniform, high-quality compound. After a defined mixing period, the compound is dropped onto a mill or extruder, where mixing is completed, and the stock sheeted out for ease of handling. Alternatively, the compound can be passed into a pelletizer. Depending on the complexity of the formulation, size of the internal mixer, and application for which the compound is intended, the mix cycle can be divided into a sequence of stages. For an all-natural rubber compound containing 50 phr carbon black, 3 phr of aromatic oil, an antioxidant system, and a semi-EV vulcanization system, a typical mix cycle will be as follows:

Stage 1 Add all-natural rubber; add a peptizer if required. Drop into mill at 165°C
Stage 2 Drop in carbon black, oils, antioxidants, zinc oxide, stearic acid, and miscellaneous pigments and discharge from the internal mixer at 160°C.
Stage 3 If required to reduce compound viscosity, pass the compound once again through the internal mixer for up to 90 seconds or 130°C.

Stage 4 Add the cure system to the compound and mix it up to a temperature not exceeding 115°C.

Computer monitoring of the internal mixer variables, such as power consumption, temperature gradients through the mixing chamber, and mix times, enables modern mixers to produce consistent high-quality compounds in large volumes. The mixed compound is then transported to either extruders for production of extruded profiles, calendars for sheeting or injection molding. Depending on the compound physical property requirements, compounds can be prepared on mills. Mill mixing takes longer, consumes larger amounts of energy, and gives smaller batch weights. The heat history of the compound is reduced, however, and this can be advantageous when processing compounds with high-performance fast acceleration systems. Two-roll mills function by shear created as the two rolls rotate at different speeds (friction ratio). This ratio of rolls speeds is variable and is set dependent on the particular type of compound. The higher the friction ratio, the greater the generated shear and intensity of mixing.

In tire factory compound production, it is always suggested that policies are in place to maximize compound shelf-life, i.e., the time from when the compound is finished mixing to when the compound is extruded or calendared into a tire component, such as tread. This is important for efficient plant operations, minimization of industrial waste, and costs. Every tire manufacturer and even every tire factory has its unique set of operating conditions. However, some principles can be considered and applied to ensure optimum shelf-life:

1. Use of peptizers for viscosity control
2. Constant viscosity natural rubbers
3. Lower Mooney viscosity polymers
4. Inclusion of retarders
5. Inventory control
6. Blending one bale of synthetic polyisoprene in place of natural rubber
7. Lower lay-down temperatures at the end of the Banbury mixing-line cooling festoon.

In addition, in a number of other raw materials, "control of use" is important and can be briefly summarized as:

1. Minimize emissions. For example, some materials will sublime, condensing on overhead pipe racks or the ceilings
2. No dust, no free powders but have them on a wax or clay carrier, or in pastille form
3. No odor
4. Non-nitrosamine
5. Banbury Injection systems, thereby avoiding free weighment
6. Pre-weigh
7. No free carbon black, i.e., use pneumatic conveyors and internal mixer injection systems thereby avoiding use of free carbon blacks
8. Where needed, scrubbers to remove regulated volatile organic compounds

6.14 SUMMARY

This chapter has reviewed both the types and the properties of elastomers, compounded with a range of filler or reinforcement systems, such as carbon black, and enhancement of filler performance by novel use of compounding ingredients such as silane coupling agents. Other issues, such as antioxidant systems and vulcanization systems, were also discussed. The role of the modern materials scientist in the tire and rubber industry is to use materials to improve current products and to develop new products. Four key parameters govern this development process:

1. Performance: The product must satisfy customer expectations.
2. Quality: The product must be durable and have a good appearance, and appropriate inspection processes must ensure consistency and uniformity.
3. Environment: Products must be environmentally friendly in manufacturing, use, and disposal.
4. Cost: The systems must provide a value to the customer.

In meeting these goals, rubber compounding is now a complex science, necessitating knowledge in advanced chemistry, physics, mathematics, and engineering.

REFERENCES

1. Rodgers B. Natural Rubber and Other Naturally Occurring Compounding Materials. In *Rubber Compounding Chemistry and Applications*, Ed. B Rodgers. CRC Press/Taylor & Francis, Boca Raton, FL. 2015.
2. *Rubber Statistical Bulletin*. 2019, July–September.
3. Roberts AD. *Natural Rubber Chemistry and Technology*. Oxford University Press, New York, NY. 1988.
4. *Rubber World*. Vol. 261, No. 1, pp. 32–48. 2019.
5. The International Standards of Quality and Packaging for Natural Rubber Grades, The Green Book, *The International Rubber Quality and Packaging Conference*, Office of the Secretariat. The Rubber Manufacturers Association, Washington, DC. January 1979.
6. American Society for Testing and Materials. *ASTM D3192. 'Natural Rubber'. Annual Book of ASTM Standards*, Vol. 09.01. American National Standards Institute. 1999.
7. International Standards Organization. *ISO 2000, Rubber Grades*. 1964.
8. Rodgers B, Halasa A. Compounding and Processing of Rubber/Rubber Blends. In *Encyclopedia of Polymer Blends*, Ed. A Isayev. Wiley. 2011.
9. Nordsiek KH. *Kautschuk Gummi Kunstsoffe 3*. 1985.
10. Ignatz-Hoover F, To BH. Vulcanization. In *Rubber Compounding Chemistry and Applications*, Ed. B Rodgers. CRC Press/Taylor & Francis, Boca Raton, FL. 2015.
11. Hong S. Antioxidants and Other Protectant Systems. In *Rubber Compounding Chemistry and Applications*, Ed. B Rodgers. CRC Press/Taylor & Francis, Boca Raton, FL. 2015.
12. American Society for Testing and Materials. *ASTM Standard D3184 - 89. 'Natural Rubber'. Annual Book of ASTM Standards*, Vol. 09.01. 1989.
13. Rodgers MB. *High Temperature Vulcanization of Industrial Rubber Products*. PhD Thesis, The Queen's University of Belfast, Northern Ireland. 1983.

14. Rodgers MB, Tracey DS, Waddell WH. *Tire Applications of Elastomers. Part 1 Treads.* Paper No H, Presented at a Meeting of the American Chemical Society Rubber Division, Grand Rapids, MI, 2004.
15. *Synthetic Rubber Production and Consumption.* ELL Technologies LLC. 2020.
16. Global Synthetic Rubber Production. 2020. https://www.statista.com/statistics/618804/total-global-natural-and-synthetic-rubber-production.html.
17. *The Synthetic Rubber Manual*, 20th Edition, International Institute of Synthetic Rubber Producers, Houston, TX. 2018.
18. Colvin H. General Purpose Elastomers. In *Rubber Compounding Chemistry and Applications*, Ed. B Rodgers. CRC Press/Taylor & Francis, Boca Raton, FL. 2015.
19. Rodgers B, Waddell W. The Science of Rubber Compounding. In *Science and Technology of Rubber.* Eds. JE Mark, B Erman, CM Roland. Academic Press. 2015.
20. Rodgers B, Tracey D, Waddell W. *Tire Application of Elastomers, 1 Treads.* Presented at a Meeting of the American Chemical Society Rubber Division, Grand Rapids, MI. 2004.
21. *ASTM D1566-04. Standard Terminology Relating to Rubber.* 2004.
22. *ASTM D3053-04. Standard Terminology Relating to Carbon Black.* 2004.

7 Radial Tire Compound Polymer Blends

7.1 INTRODUCTION

In terms of global science and engineering trends, demographics and development, four fundamental enabling technologies have been suggested that will provide the foundation for the future, namely i) information technology, including artificial intelligence, communications, and automation, ii) enhanced concrete as a building material, iii) energy sources, both conventional and alternative, and iv) polymer blending. The observation on polymer blending is based on the assumption that all possible polymers, and the monomers to produce them, have been invented, and thus the best means to improve material properties is by blending existing polymeric materials. In the field of tire engineering, blending of rubbers therefore merits important consideration.

Blending of elastomers in tire compounds is done for two principal reasons; first, improvement of the mechanical properties of the base compounded elastomer, and second, improved processing behavior of the compounded polymer (1). Processing would include mixing, extrusion, and calendaring of rubber compounds, which will also contain oils, carbon blacks, and other organic and inorganic chemicals. Technical properties, such as mechanical and dynamic characteristics, are impacted by the selection of elastomers in a blend, distribution of vulcanization system ingredients in the blend, and filler distribution between elastomer phases. Filler distribution will be affected by the type of elastomers, the molecular weight of each of the different polymers, the mixing method, and polymer–filler interaction. There are thus three areas of importance to the tire engineer:

1. Elastomer blending and compounding,
2. Secondary elastomer blend compound systems, such as polymeric resins, forming interpenetrating networks (IPN), and
3. Elastomer blends and tire performance.

In complex structures, such as tires, elastomers typically tend to be blends to obtain specific properties, so that the ultimate product may function according to the required mission profile. There are a number of compounding references that the reader may find of value and which are provided at the end of this discussion (2, 3, 4).

7.2 APPLIED POLYMER TECHNOLOGY

There are also a number of definitions which first merit discussion. Glass transition temperature or Tg defines the temperature at which an elastomer undergoes a

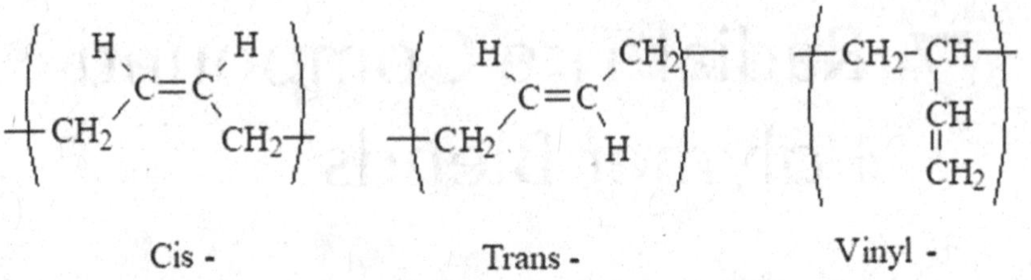

FIGURE 7.1 Possible Configurations for Butadiene in Solution SBR and BR (1, 2)

transition from a rubbery to a glassy state at the molecular level. This transition is due to a cessation of molecular motion as the temperature drops. Increase in the Tg, also sometimes referred to as the second-order transition temperature, leads to an increase in compound hysteretic properties, and, in tires, an improvement in tire tread traction characteristics. A decrease in Tg leads to a decrease in compound heat build-up or hysteresis and, typically, to improved resistance to abrasion and better rolling resistance.

Similarly, the microstructure of an elastomer can affect the mechanical properties of the compounded elastomer. For solution SBR used in a tire tread compound, an increase in the vinyl butadiene (Figure 7.1) level will increase the Tg, improve tire wet traction performance, and result in a decrease in resistance to abrasion. Increases in the concentrations of *cis*-and *trans*-isomers of butadiene in SBR, with a corresponding decrease in the vinyl-1-butadiene isomer, will improve compound abrasion resistance. The amount of styrene in SBR will affect tire traction. Higher levels of styrene tend to give improvements in tire traction and tire–vehicle handling properties, due to an increase in the dynamic modulus (G'). The ratios of styrene, *cis*-butadiene, *trans*-butadiene, and vinyl-butadiene determine the ultimate Tg of the polymer. For example, the higher the *cis*- or *trans*-butadiene level, the lower the Tg of either SBR or BR (2).

The microstructure of BR can have a significant effect on the elastomer's performance. For example, lithium-catalyzed solution polymers, containing approximately 36% *cis*-butadiene, tend to process easily in tire manufacturing plants, whereas high *cis*-polymers (*cis*- contents greater than 90%) produced with titanium or nickel catalysts tend to be more difficult to process in tire factories (Table 7.1). For

TABLE 7.1
Polybutadiene Microstructure Typical Values
(In Absence of a Modifier)

Catalyst Base Metal	Cis-%	Trans-%	Vinyl-%
Li	35	56	10
Ti	91–94	2–4	4
Co	1	2	2
Nd	1	1	1
Ni	96–98	0–1	2–4

example, gauge control of extruded components may be more difficult though such compounded polymers have been reported to show better abrasion resistance when compared with the low-*cis* polymer.

An example of how polymer microstructure and polymer Tg impact performance occurs when vinyl-butadiene is increased from 10% to 50% in BR (Table 7.2). The Tg increases from −90 to −60°C, with a corresponding shift in the *tan δ* curve. Traction performance improved significantly but tread wear and rolling resistance performances declined.

The concept of integral rubber, more frequently referred to as dual or multi-Tg polymers, emerged from this fundamental work (6, 7).

Rather than blending an SBR with a BR or polyisoprene, for example, copolymers or terpolymers based on styrene and butadiene, butadiene and isoprene, styrene and isoprene, or styrene, butadiene and isoprene, can be prepared having a dual Tg. Using *n*-butyllithium catalyst and a polar modifier such as N,N,N′,N′-tetramethylethylenediamine (TMEDA), a range of polymer dual-glass transition temperatures can be obtained (Table 7.3).

TABLE 7.2

Effect of Tread Polymer Butadiene Vinyl- Level on Tire Performance (5)

	10%	50%
Percent vinyl- Butadiene		
Glass transition temperature (Polymer Tg, °C)	−90	−60
Tire properties		
• Wet traction	100[1]	120
• Rolling resistance	100	95
• Tread wear	100	90

[1] Ratings: control at 100; higher rating is better

TABLE 7.3

Dual Tg IBR, SBR, and SIBR Elastomers (1)

Polymer	Microstructure Tg (with TMEDA Polar Modifier)				
	Styrene	Butadiene	Isoprene	Lower Tg (°C)	Upper Tg (°C)
1	—	100	—	−98	−50
2	—	50	50	−80	−30
3	—	80	20	−85	−54
4	45	45	10	−75	−25
5	40	40	20	−78	−18
6	35	35	30	−78	−10

In addition to styrene, TMEDA concentration in the polymerization process will control the levels of vinyl-butadiene and 3,4-isoprene, and this can determine the upper Tg, whereas the concentration of *cis*-butadiene will control the lower Tg. This work led to the development of the styrene-isoprene-butadiene terpolymer, which proved to be very effective in obtaining a balance of increased abrasion resistance, traction, vehicle-handling qualities, and decreased rolling resistance (2, 7) . Trends can then be established, illustrating how microstructure can be modified to achieve a specific combination of Tgs. This, in turn, allows specific tuning of the compounded rubber, for example, the tire tread compound necessary to meet a special set of tire traction and rolling resistance performance criteria.

Thus, optimizing the polymer microstructure to obtain the final compound properties can offer significant tire manufacturing efficiencies. However, tire companies that are not back-integrated in the supply of synthetic rubbers must use conventional polymer blending methods to achieve the required mechanical properties of the compounds.

7.3   TIRE COMPONENTS

The pneumatic tire consists of two basic areas: the tread area, which is responsible for ground contact, and the casing, the function of which is to support load and transmit power to the tread area. Each of these areas has several components with markedly different properties and which serve specific and unique functions, and all of which must interact together to achieve the designed performance. The role of the rubber compounds, which are used in these basic components, is three-fold:

1. To provide the contact area between the vehicle and the road surface.
2. To provide the cohesive material that holds the tire together so it acts as an integral unit, and
3. To provide protection for the strength bearing components, i.e., the textiles, steel beads, and steel breakers in steel-belted radial tires.

Figure 7.2 shows a cross-section of a pneumatic tire. The tread is designed and compounded to achieve high abrasion resistance and traction, low rolling resistance, and protection of the casing. It is often divided into two subcomponents to tune performance, namely the outer tread, for surface contact, and the under-tread, which serves as a transition component and reduces tire rolling resistance through decreased hysteresis (Chapter 2).

The elastomers and elastomer blends used in tire compounds, and particularly tread compounds, are equally important to the structural design parameters of the tire (Table 7.4). A properly designed tread compound will ensure the tire can meet its performance targets. Tread compounding materials fall into one of five general categories: polymers, fillers, protectants, vulcanization system, and a variety of special-purpose additives. Elastomers for tread compounds are typically natural rubber (NR), styrene-butadiene rubber (SBR), polybutadiene (BR), and, in some instances, isobutylene-based polymers for winter and special-performance tires.

For the mechanical strength requirements of tires, NR is better than SBR or BR, enabling better resistance to tread chipping, chunking, and cutting, lower tire

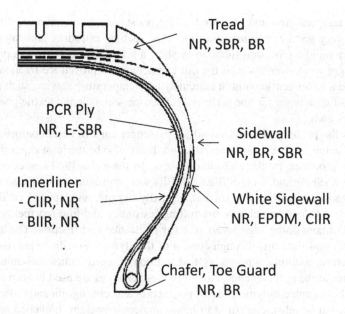

FIGURE 7.2 Tire Construction

TABLE 7.4
Typical Tread Compound Formulations and Quantities of Raw Materials (8)

Compound	Standard Economy Line Tire Tread (phr)	High Performance (phr)	Low Rolling Resistance (PHR)	Radial Medium Truck (PHR)
Polymer	SBR 50–100	VSBR 20–100	NR 0–40	NR 60–100
	BR 0–50	BR 0–20	SSBR 60–100	BR 0–40
Carbon black	N234	N121. N134	N299	N121
Grades	N299	N234	N351	N134
	N 399	N343	N343	N220
Carbon black load	70–90	80–90	40–60	40–60
Silica	80	80	80	80
Oil (incl OE in polymers)	30–60	25–45	0–30	0–20
Antioxidants Antiozonants	1–4	1–4	1–4	1–4
Processing aids	1–10	1–10	1–10	1–10
ZnO	3	3	3	3
Stearic acid	1–2	1–2	1–2	1–2
Sulfur	1	1	1	1
Accelerator types	CBS DPG	CBS. DPG	CBS. DPG	TBBS
Loading	1–2	1–2	1–2	1–2

operating temperatures, and improved rolling resistance. Although more difficult to mix, it is very good in downstream processes, such as extruding and calendaring, and for tire building. Natural rubber may show a faster wear condition in passenger car and light truck over-the-road tires. It can achieve improved ice traction performance and a lower brittle point in extremely low temperature service, such as down toward −40°C. It is, by far, the preferred rubber for wire belt and textile ply/breaker coat compounds.

SBR is the primary elastomer used for passenger car tire tread compounds, due mainly to better wear resistance and traction. It can also be the least expensive polymer when produced by the emulsion process. In the early 1980's, specialty solution SBRs (SSBR) and vinyl SSBRs (VSBR) were introduced (3). These rubbers, polymerized in a hydrocarbon solvent, typically hexane, with special catalyst systems, achieved improvements in tire rolling resistance and lowered fuel consumption, while maintaining high levels of wear resistance and traction. The specialty SSBRs accomplished this through control of the styrene–butadiene polymer backbone structure, coupling of polymers, and addition of reactive chemical units or coupling agents at the polymer ends. Not all of these changes are used in each specialty polymer, but coupled polymers are in production and can significantly affect polymer molecular weights, compared to higher-molecular-weight, branched polymers, which may be difficult to produce and may not perform as well in tires. Many of the polymers are available in the oil/black masterbatch (premixed blends), as is emulsion SBR. SBR, whether as emulsion or solution, is still not as resilient as NR or BR, limiting its applications in tires.

BR can be produced primarily by hydrocarbon solvent processes. It is the butadiene part of SBR that is controlled in specialty SSBRs. Originally, BR was used in tires due to its better abrasion resistance. It has high *cis*-butadiene content and high linearity structure, and this also results in lower hysteresis and better fatigue resistance. BR is more difficult to process in the factory, particularly in milling operations. It does not possess the level of tack of NR and makes tire building more difficult. BR is also lower in resistance to tearing and cutting.

Chlorobutyl and bromobutyl rubber (CIIR, BIIR) can be used in tire inner liners and white sidewalls. These elastomers allow improvements in tire air retention due to lower air permeability, as well as improved age resistance and flex fatigue resistance. The chlorine and bromine halogenated versions of isobutylene isoprene copolymers or butyl rubber (IIR) are both used to improve compatibility with the other commodity tire rubbers. It is a near-saturated polymer, with few sites available for conventional sulfur curing. IIR shows high hysteresis at nominal tire operating temperatures and thus acts as a superior cushioning rubber. In the 1960s, a tire using this rubber in the tread was produced by tire manufacturers to achieve improved vehicle handling performance. Vehicle–tire handling characteristics describe properties such as steering precision, torque alignment, cornering moment, stability, vibration, comfort, and noise generation. It is also the preferred rubber for tire curing press bladders.

Compared to halobutyl (HIIR), ethylene propylene rubber (EPR) has a fully saturated backbone and has only unsaturation points available for vulcanization crosslinking in very low percentages of the pendant diene modifier (EPDM). It has good aging resistance and resistance to fatigue even when heavily loaded with fillers, and is utilized in passenger car tire white sidewalls.

Another example of how polymer blends are used is in truck tire sidewalls, which must possess not only high strength but also excellent aging and fatigue resistance in order to survive multiple retreading, i.e., vulcanizing a new tread to the worn original tire casing. The original casing (carcass) of a radial truck tire has been known to last for up to 750,000 miles having had two retreads, whereas the original tread on this casing delivered only 70,000–350,000 miles.

In passenger car tires, as many as four different polymers may be used for the tread compound, totaling 100 phr e.g., 25 phr emulsion SBR, 25 phr SSBR, 30 phr BR, and 20 phr NR. If solution SBR categories falling with a 10°C Tg range can be considered as specialty SSBRs, at least nine groups of specialty SSBRs are produced. SBR is available as oil and oil–carbon black masterbatches (oil and/or carbon black added to the rubber by the rubber producers), which may use a different SBR (usually higher molecular weight) than if it were produced without masterbatch. Mooney viscosity of oil-extended elastomers is sometimes called the bale mooney, typically around 55 Mooney units (MU). The base mooney is the original polymer mooney viscosity, typically as high as 120 MU, before oil addition lowers it to achieve the final bale mooney.

7.4 BLENDING ELASTOMERS

Compounded elastomers have many unique characteristics not found in other materials, such as high elasticity, abrasion resistance, and dampening properties. A compound used in a tire tread, for example, consists of five major components, namely the polymer system, filler or reinforcement system, processing aids, antioxidants, and the vulcanization system. Frequently, more than one polymer is used in a compound used in tire treads or sidewalls. There are some additional definitions which should be reviewed before discussing polymer blending. Compatibility of two elastomers is where the polymers are immiscible, but, in combination, provide properties or show characteristics that are more useful than the properties of the original polymers individually. Incompatible polymer blends typically show a dual Tg.

Miscible blends are obtained when, at the molecular level, the two polymers are compatible. For example, blends of two highly polar polymers, such as PVC and NBR, are truly miscible elastomeric systems. Examples of elastomers that exhibit phase separation upon heating include natural rubber and high vinyl solution SBR. This is important at high curing temperatures (180°C), where rapid phase separation can occur before the onset of crosslinking. However, at curing temperatures, natural rubber and conventional emulsion SBR forms a single phase. A number of terms can now be defined:

Miscible: Polymer blends show a single Tg and single-phase morphology.
Semi-miscible: Miscibility varies with temperature and concentration.
Immiscible: Polymers show two-phase morphology with a distinct domain
    size and shape
Compatible: Polymers are not miscible at the molecular level.

Incompatible Showing two or more intermediate Tgs and two-phase morphology or more.

A wide variety of compounds contain polymer blends. This can facilitate improved
fatigue life, groove cracking resistance, hysteresis, and often increased adhesion and
abrasion resistance. Blending also improves extrusion, mold-ability, and calendaring
characteristics of a compound. For example, addition of natural rubber to a compound with a high synthetic polymer level will improve tire buildability, as well as
improving resistance to tear strength of the compound. This can be important in
curing and de-molding of tires with high traction, low rolling resistance treads that
often have low hot tear strength when the press opens.

For two substances of different chemical composition to be miscible, the Gibbs
Free Energy of Mixing, ΔG, must be negative (Equation 7.1). Gibbs Free Energy
is a function of enthalpy of mixing (ΔH), entropy of mixing (ΔS), and the absolute
temperature (T) where;

$$\Delta G = \Delta H - T\Delta S \tag{7.1}$$

Miscible polymers tend to show hydrogen bonding or van der Waals forces between
the chains. However, phase separation can occur should the temperature rise above
the lower critical solution temperature. Polymers generally do not have an upper
critical solution temperature, although lower-molecular-weight oligomers do. A final
point to note is that truly miscible polymers will show a single Tg. Two-phase heterogeneous systems are obtained both above the lower critical solution temperature
and below the upper critical solution temperature. The critical solution temperatures
also vary according to the polymer ratios present. The phase morphology which two
immiscible elastomers assume can take a variety of forms, depending on the viscosity or molecular weights, solubility, and polarity of the polymer chain. Examples
might best be described as:

1. Rubber particles within a plastic matrix
2. Plastic particles within a rubber matrix
3. Plastic networks in a rubber matrix
4. Rubber rods in a plastic matrix,
5. Plastic rods in a rubber matrix
6. Plastic-rubber alternating sheets

Many more alternatives could be developed, with phase inversion between the polymer
phases being dependent on polymer concentration, molecular weight, or mixing method.

BR and NR are blended in tire sidewalls to improve fatigue resistance, cut growth
resistance, and abrasion resistance to minimize the effect of curb scuffing. A model
tire sidewall compound formulation is shown in Table 3.6. Elastomer blends in tire
compounds can be miscible, compatible, or incompatible. The control of this phase
morphology can become very important so as to prevent any deterioration in the tire
tread wear performance, rolling resistance, traction and handling qualities, or loss in
component-to-component adhesion.

Miscibility can be predicted from the Flory-Huggins theory, using interaction parameters, and, in many instances, are in close agreement with experimental data. Some empirical guidelines on elastomer miscibility can thus be proposed as follows:

1. In general, the Tg of miscible blends is intermediate between those of the component polymers and may be a function of the weight percentage or the ratios of the constituent polymers. Miscible blends have one Tg, immiscible blends have two Tgs.
2. For SBR and BR blends, when the styrene level in SBR increases beyond 30%, immiscible compositions are obtained.
3. The miscibility of polyisoprene and SBR increases with increase of 1,2-isoprene.
4. 1,4-IR and 3,4-IR become immiscible when the 3,4isoprene content exceeds 50%. Miscibility can be improved by lowering the molecular weight of the vinyl IR polymer
5. For polyisoprene–polybutadiene blends
   a. Miscibility increases with BR 1,2-butadiene content
   b. IR high-vinyl BR blends can have a broad graduated Tg, which can sometimes be described as semi-miscible systems

Polymer blends in tire tread compounds have been reported to offer a number of performance advantages, which can be summarized as follows:

- BR and NR
  - improves heat stability,
  - decreases tensile strength and 300% modulus,
  - improves resistance to fatigue
  - increases resistance to abrasion and tread wear, and
  - increases elasticity.
- BR and SBR
  - an increase in SBR level improves traction (due to an increase in Tg),
  - SBR increase improves tire–vehicle handling qualities, due to the increase in dynamic modulus (E', G')
  - increase in percentage BR improves wear performance,
  - blends of NR, SBR, BR can be optimized, depending on the tire design criteria, for factors such as aligning torque, steering precision, high-speed performance, etc.

7.5 PROCESSING OF ELASTOMER BLENDS

Generally, elastomers can be blended in one of five ways: i) latex blending, ii) solution blending, iii) combined solution/latex blending, iv) mechanical blending, and v) mechano-chemical blending. Of these methods, latex blending, solution blending, and mechanical blending are the most important (1, 4).

Latex blending gives a high-quality dispersion of component materials, with the blend potentially having lower hysteretic properties and lower solvent swelling.

Examples of where such blending is found would be in NR/SBR latex blends. Solution blending of elastomers is normally done during the solution polymerization process. At the same time, carbon black and oils can be added to produce carbon black masterbatches.

Solution blending of incompatible polymers is difficult due to phase separation. Furthermore, rapid solvent evaporation is required to minimize heterogeneity, with these two problems restricting the applications of this technique. Combined solution and latex blending have an advantage in, for example, solutions of polymers mixed with aqueous suspensions. Carbon black migrates into the organic phase, thereby giving a highly homogeneous dispersion.

Mechanical mixing of polymers, carried out on either an open mill or in a Banbury internal mixer, has advantages in terms of increased efficiency, increased productivity, and lower cost. However, NR, in many instances, must first undergo an initial breakdown.

Polymer blends can be characterized in a number of ways.

1. Microscopic: characterize the compatibility of polymers at the gross scale
2. Electron microscopy
3. Solubility measurements by swelling methods
4. Thermal and thermo-mechanical methods such as DSC
5. Dielectric relaxation
6. Infrared spectroscopy
7. NMR
8. X-ray analysis
9. Gel permeation chromatography
10. Refractive index
11. Viscosity
12. Light scattering, and
13. Rheological properties

A number of observations can therefore be highlighted:

1. Phase micrographs can show discrete domains of polymers in blends. The dispersed zones range from 1.0 up to 30 µm in diameter, but more frequently 1.0–5.0 µm.
2. The area of a polymer domain approximates to the proportion of the polymer volume in the blend.
3. The discrete zones are the secondary polymer dispersed in the continuous phase of the major polymer. Phase inversion can occur at the 50/50 blend ratio.
4. The mid-point is characterized by creation of an interpenetrating network.
5. Components of the lowest viscosity form the continuous phase (e.g., NR/CR blends where NR is the continuous phase up to a 25/75 blend ratio).
6. An increase in mix time and temperature will decrease the domain size.
7. IR/SBR and IR/PBD blends are heterogeneous. SBR/BR blends are homogeneous.

8. Ozone resistance improves as the fineness of the dispersion increases.
9. When BR has a high viscosity, blending is poor; there are large BR domains.
10. If the NR and BR polymers used have equivalent viscosities, then the blend will tend to be more homogeneous. This is also true of NR/SBR blends.
11. At high carbon black loadings in BR, the polymer is forced into the dispersed phase. The blend morphology is secondary to the impact of the carbon black dispersion.

In summary, elastomer blends are essentially micro-heterogeneous systems. The continuous phase is either the polymer at the highest concentration or the polymer with the lowest viscosity. Phase inversion occurs at a ratio of typically 50/50 and zone sizes are in the order of 1–5 μm. Micro-crack growth termination is enhanced by smaller domains, but poor dispersion of soluble compounding ingredients can have a detrimental effect on compound mechanical properties.

7.6 NATURAL RUBBER–BROMOBUTYL BLENDS

Natural rubber is added to bromobutyl to improve calendaring and to improve green strength (8). For example, high compounded green strength is important in resisting tire ply cords pulling through a tire inner liner when the tire is being built and shaped. Although bromobutyl and natural rubber have different backbone structures, different degrees of functionality, and different vulcanization chemistries, they can be blended in various amounts to attain a desired combination of properties. This is due to the versatility of bromobutyl rubber vulcanization. Not only can bromobutyl rubber vulcanize independently from the natural rubber–sulfur cure system by merely crosslinking with zinc oxide, but it also uses sulfur efficiently in the presence of natural rubber to promote co-vulcanization (9, 10).

The addition of bromobutyl rubber to a predominantly NR compound reduces the permeability to gases and water vapor, increases the heat and flex crack resistance, and increases vibration dampening. Blending NR into bromobutyl rubber increases cured adhesion to general-purpose rubber compounds as well as increasing building tack and green strength. In blends of the two rubbers, the physical properties, in particular tensile strength, usually show a significant change compared with the properties of the individual rubber compounds (11). This infers either that one rubber disturbs the cured network of the other, or that the two immiscible phases are curing separately. With bromobutyl–natural rubber blends, tensile strength values can be equal to or greater than the average of the two homo-polymers, which implies crosslinking between the elastomer phases. A thiazole–sulfur vulcanization system generally exhibits the best balance of properties with blends containing eighty percent or more of bromobutyl. Below seventy percent, an alkylphenol disulfide cure system is suggested.

Permeation of a gas or permeability through a membrane such as a tire inner liner consists of three distinct processes. First, the gas molecules must dissolve on one side of the membrane, then diffuse across the membrane or liner to the opposite side of lower gas concentration, the rate being dependent on the size of the diffusion gradient, and then evaporate or disperse in the adjacent tire component or other medium.

The rate of diffusion of oxygen and nitrogen through a butyl, chlorobutyl, or bromo-butyl rubber membrane is a function of a number of parameters. The size of oxy-gen and nitrogen molecules is 2.9 and 3.1 angstroms, respectively, with the smaller molecule expected to diffuse more rapidly. This is observed in practice. Amerongen (12, 13), in calculating energy of activation of gas diffusion, also noted that diffu-sion coefficients can increase between 1.2 and 3 times with each 10°C increase in temperature. The activation energy of diffusion was reported to be 50.6 and 49.8 KJ/mol for nitrogen and oxygen, respectively, with frequency constants, analogous to the Arrhenius constants, of 180 and 142, respectively. From thermodynamics, this clearly suggests that oxygen diffuses more readily than nitrogen, as is observed. The permeability of gases through compounded bromobutyl or compounded bromobu-tyl/NR blends is defined as the amount of gas at standard temperature and pressure (STP) permeating through the compound sheet of area, A, thickness or gauge, l, at a defined time interval, typically 1 day, with a pressure differential across the com-pound sheet. The permeability coefficient, P, can therefore be expressed in units of cc × mm/(m^2.day.mmHg). The permeability coefficient, P, is related to the diffusion coefficient, D, and the solubility constant, S, by the expression:

$$P = DS \tag{7.2}$$

Diffusion is dependent on the activation energy of diffusion and thus can be esti-mated through use of an Arrhenius expression:

$$D = D_o \, e^{-Ed/RT} \tag{7.3}$$

where E_d is the activation energy of diffusion. The value of E_d is related to the poly-mer and the specific gas and will vary depending on the Tg. The solubility constant shows a similar relationship,

$$S = S_o \, e^{-\Delta Hs/RT} \tag{7.4}$$

where ΔHs is the heat of sorption. Combining equations 7.3 and 7.3 gives:

$$P = P_o \, e^{-Ed/RT} \tag{7.5}$$

where Ep can be referred to as the activation energy of permeation. The permeability coefficient for various compounded NR and bromobutyl blends is shown in Table 7.5.

It is difficult to alter the scorch properties of bromobutyl/general-purpose rubber blends with classical retarders such as N-cyclohexylthiophthalimide. For instance, salicylic acid is a vulcanization retarder for general-purpose rubbers (GPR), but an accelerator for bromobutyl rubber. Alkaline retarders used with bromobutyl rubber are often accelerators for general- purpose rubbers. Therefore, the scorch time of blends is best set by choosing the appropriate vulcanization system and the com-pound viscosity (as it influences the heat history of the compound during process-ing). Magnesium oxide is frequently used at between 0.1 and 0.5 phr as a scorch

TABLE 7.5

Model Formulations for an Automobile Tire Inner Liner and Summary of Properties (1, 9)

Compound	Test Method	Units	1	2	3	4	5
EXXON™ bromobutyl grade 2222		PHR (1)	100	90	80	70	60
Natural rubber		PHR	0	10	20	30	40
Carbon black N660		PHR	60	60 60	60	60	
Naphthenic oil		PHR	8	8	8	8	8
Aromatic and aliphatic hydrocarbon I		PHR	7	7	7	7	7
Phenolic tackifying resin		PHR	4	4	4	4	4
Magnesium oxide		PHR	0.15	0.0	0.0	0.0	0.0
Stearic acid	PHR	2	2	2	2	2	2
Zinc oxide		PHR	1	1	1	1	1
Sulfur		PHR	0.5	0.5	0.5	0.5	0.5
Mercaptobenzothiazyl disulfide (Ml)		PHR	1.5	1.5	1.5	1.5	1.5
Total		(PHR)	184.15	184	184	184	184
Properties	Test Method	Units					
Mooney viscosity (Ml)	ASTM D1646	MU. 100°C	53.4	50.6	46.4	43.6	42.5
Mooney scorch	ASTM D1646 @	125°C					
Time to 5pt rise		Minutes	29.5	22	19.2	16.9	16.8
Rheometer (MDR)	ASTM D5289	160°C; 0.5° arc; 60 minutes					
Mh-MI (delta torque)		dNm	3.7	3.3	3.8	4.4	5.1
Tc90 (time to 90% torque increase)		Minutes	12.6	7.6	8.4	8.8	9.5
Tensile strength	ASTM D412	MPa					
9.9	9.8	9.7	10.6	11.6			
Elongation at break		%	905	908	809	774	715
Modulus 300%		MPa	2.5	2.4	2.8	3.1	3.7
Tear strength (Die E)	ASTM D624	N	98.1	88.5	87.5	94.7	121.4
Tear strength (Die B)		KN/m	51.2	46.1	45.7	49.6	62
Mocon oxygen transmission		0.7	1	1.2	1.4	1.7	
cc"mm/(m²-day-mmHg) @ 60°C							

Note 1. ExxonMobil Chemical Data. See 'www.butylrubber.com' for a more comprehensive list of properties Compound

inhibitor in compounds containing 100 phr content bromobutyl rubber or bromobutyl–NR blends.

7.7 HALOBUTYL–BUTYL BLENDS

With bromobutyl–butyl rubber blends, elastomer structures are essentially the same, but the different reactive functionalities provide different vulcanization chemistries. Since bromobutyl rubber has greater cure reactivity, accelerators that will over-cure the bromobutyl rubber phase should be avoided. Briefly:

1. Low level blending of butyl rubber into bromobutyl rubber increases the scorch time and improves the heat resistance of the compound.
2. Blends with up to 20 phr of butyl rubber exhibit bromobutyl rubber-like adhesion.
3. Butyl rubber levels in excess of 20 phr sharply reduce adhesion to general-purpose rubber compound substrates. Blending up to thirty percent butyl into bromobutyl rubber slows the cure rate and provides a small benefit in lower modulus increase and higher retained elongation after heat aging.

Because butyl has the same permeability coefficient as bromobutyl rubber, blending it into a bromobutyl rubber compound does not alter the compound permeability as, for example, NR would. In bromobutyl–chlorobutyl rubber blends, both elastomers have the polyisobutylene backbone and halogen reactive functionality. These polymers, being miscible at the molecular level, constitute an ideal system for co-vulcanization. Bromobutyl and chlorobutyl can be used interchangeably without significant effect on the state of cure as measured by extension modulus, tensile strength, and cure rheometer torque development. Bromobutyl will increase the cure rate of a blend with chlorobutyl. However, where bromobutyl is the major part of the blends, chlorobutyl does not reduce scorch tendencies because the more reactive halogen unit will dominate.

7.8 BROMOBUTYL–GPR BLENDS

Like NR, SBR can be blended in all proportions with bromobutyl rubber. However, SBR is less desirable for blending than natural rubber due to its low tack and green strength properties. In addition, resistance to heat, flex fatigue resistance, and weathering are poorer with SBR blends than with NR blends. Suggested cure systems are the same as those for bromobutyl–NR blends (11).

Ethylene propylene diene terpolymers (EPDM) can be used to improve the ozone resistance of bromobutyl–NR binary polymer blends, eliminating the need for chemical antiozonants. An addition of 10 phr of EPDM (with a high ethylidene norbornene, ENB, content of 9%) to a 50/50 bromobutyl rubber/NR blend results in a compound with good static and dynamic ozone resistance. EPDM with a 5.7% ENB level is another suggested grade of polymer. Such blends have significant industrial significance as they are frequently used to prepare white sidewalls and white lettering on tires. In these blends, natural rubber contributes adhesion to adjacent components and lower energy loss, EPDM imparts static ozone resistance, and bromobutyl rubber provides flex and dynamic ozone resistance. EPDM rubbers with the highest degree of unsaturation are suggested for high state of vulcanization in general-purpose rubber–bromobutyl blends. Bromobutyl rubber and either EP or EPDM rubber can be blended in all proportions and can be co-vulcanized using either zinc oxide/sulfur cure systems or peroxide-based cure systems.

7.9 DISTRIBUTION OF COMPOUNDING INGREDIENTS: INSOLUBLE CHEMICALS

When NR and BR are blended, carbon black normally locates preferentially in the BR phase This distribution can achieve better compound performance, such as improved tensile strength, tear strength, and hysteretic properties. Carbon black particles may migrate from the NR masterbatch phase to the BR gum phase during mixing, when the compound is at elevated temperatures and this transfer continues until phase saturation occurs. Under these conditions, steric bulk of the carbon black aggregate, polymer viscosities, and mixing temperatures would be the principal factors affecting particle distribution in the final compound.

The incorporation of carbon black into 50/50 elastomer preblends shows black affinity decreases in the order: BR, SBR, CR, NBR, EPDM and butyl rubber. Poor carbon black dispersion causes an increase in hysteresis and a decrease in fatigue resistance. The effect of volume loading of a filler, such as carbon black, on the properties of the vulcanizate depends on whether the elastomer is strain-crystallizing or not.

The tensile strength of filled elastomers over the tensile strength of the gum rubber is much greater for SBR. A greater increase in tensile strength is attainable for non-crystallizable polymers. Filler distribution is also affected by filler type. Elaborating:

1. Reinforcing fillers such as carbon black and silica have a critical loading factor that is different for each polymer and filler type. For example, the optimum loading for SAF grade carbon blacks in natural rubber will be in the range 45–60 phr, whereas, for highly dispersible silicas in solution SBR, it will be in the range of 75–85 phr.
2. Non-reinforcing fillers, such as clay and calcium carbonate, may be enhanced with surface treatments such as carboxylated BR, calcium stearate, or an organo-silane coupling agent.
3. With all types of fillers, distribution is important for wear resistance and tear strength.

As already noted, a higher distribution of carbon black in the BR phase will improve vulcanizate properties. As an empirical guideline, tire tread wear and compound tensile strength are improved while tear strength improves and then decreases when a higher percentage of carbon black is located in the BR phase (Figure 7.3). Similarly, as an empirical guideline, Figure 7.4 illustrates that hysteresis is also improved as the percentage of carbon black in the BR phase is increased.

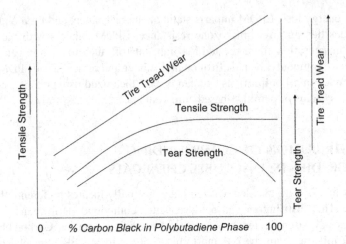

FIGURE 7.3 Carbon Black Dispersion and General Trends in Elastomer Compound Properties (1, 4, 14)

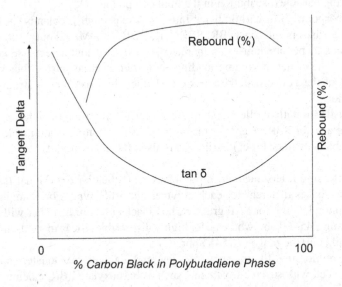

FIGURE 7.4 Carbon Black Dispersion and General Trends in Compound Hysteresis (1, 4, 14)

The set-up of the factory mix cycle for sidewall compounds, which are typically blends of NR and BR, are very dependent on the mix. A mixing sequence allowing high levels of dispersion in both NR and BR phases will ensure the lowest possible hysteresis as well as good abrasion and scuff resistance.

These empirical trends suggest that an optimum distribution of 60% of the black in the BR phase will give the best hysteretic characteristics in a compound. Factory compound mix cycles are therefore set up accordingly.

As for NR–SBR masterbatch blends, carbon black will migrate to the SBR gum phase. At equilibrium, the amount of carbon black in the NR phase will depend on the particle size. For example, the amount of carbon black remaining in the NR phase can drop from 41% for NSSO to only 27% when using N134. Other factors or empirical guidelines affecting the distribution of fillers include:

1. The surface polarity of carbon black influences its distribution in elastomer blends.
2. Silica will tend to accumulate in the NR phase.
3. Filler phase distribution is influenced by the point of addition to the mix cycle.
4. The use of coupling agents will potentially bind a filler into one phase.
5. Bound rubber (NR/CB) tends to form more readily in the carbon black-loaded NR phase than in the BR phase

Carbon black critical loading factor or optimum loading level will be a function of the specific carbon black grade and the specific grade of polymers in the rubber compound blend, and is therefore unique for each polymer and filler system.

7.10   DISTRIBUTION OF COMPOUNDING INGREDIENTS: SOLUBLE CHEMICALS

The distribution of soluble compounding ingredients, especially curatives, may significantly influence the performance of the vulcanized product. Diffusion of common vulcanizing ingredients, such as sulfur, tetramethylthiuram disulfide (TMTD), mercaptobenzothiazole disulfide (MBTS), and diphenyl guanidine (DPG) occurs from compounded highly saturated elastomers, such as IIR and EPDM, to highly unsaturated elastomers, such as NR and SBR.

The migration of curatives was observed to happen very quickly (for example, 3 seconds at 150°C). A diffusion gradient may be produced between the dissimilar elastomers well before any vulcanization occurs. The practical consequences of the solubility and diffusion differences among elastomer blends are curative imbalance between the component phases and associated over- and under-cure.

Methods to prevent this diffusion gradient include preparing a masterbatch of dithiocarbamate accelerators in the polymer (IIR, EPDM) and grafting accelerators onto the polymer (EPDM). One advantage of a cure system mismatch is improved ozone crack resistance. In any case, cure system location control is critical. Figure 7.5 shows the diffusion of sulfur from NR to SBR. Two rubber samples are placed in contact with each other and heated. The sulfur concentration is then measured at increasing distances from the interface. In this case, considerable transfer occurs, and significant concentrations build up in the SBR phase. In contrast, the level of sulfur migration is much less from NR to butyl rubber. The solubility of sulfur in butyl is much less than the solubility of sulfur in NR (Figure 7.6). The diffusion of MBTS is similar to sulfur diffusion, but it occurs more slowly due to the large size of the MBTS molecule. MBTS is also less soluble than sulfur in each rubber type (Figure 7.7).

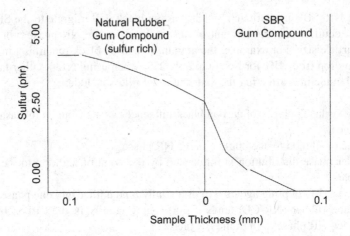

FIGURE 7.5 Illustration of the Diffusion of Sulfur From NR to SBR at 150°C (1, 4) (Adapted from Corish, Elastomer Blends, 1994, Rodgers et al., 2011)

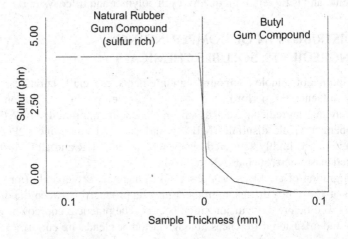

FIGURE 7.6 Simplified Illustration of the Diffusion of Sulfur From Natural Rubber to Butyl Rubber at 150°C (1, 4) (Adapted from Corish, Elastomer Blends, 1994)

To summarize the distribution of soluble compound chemicals in polymer blends:

1. Sulfur diffuses from NR to SBR until an equilibrium is reached.
2. Little diffusion of sulfur occurs from NR to butyl rubber, due to the low solubility in butyl rubber.
3. The diffusion coefficient is lower for MBTS and t-butyl-2-benzothiazole sulfenamide (TBBS) due to:
   a. the larger steric bulk of its molecule compared to sulfur
   b. the aromatic nature of MBTS and TBBS.
4. The cured properties imbalance may not be equal to that of homogeneous polymers due to the level of curative imbalance between the phases

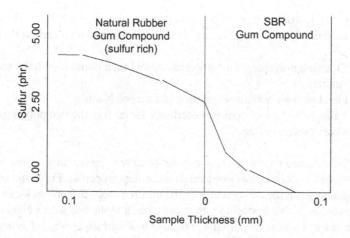

FIGURE 7.7 Illustration of the Diffusion of MBTS From SBR to BR at 150°C (1, 4) (Adapted from Corish, Elastomer Blends, 1994, Rodgers et al., 2011)

5. Oil migration is also dependent on the polymer matrix
6. Antioxidants and antiozonants diffuse to equilibrate themselves in the tire.
7. 1.3-Dimethylbutyl-N′-phenyl-*p*-phenylene diamine (6PPD), which is an important antiozonant used extensively in tires, migrates preferentially during curing as follows:
   a. EPDM to SBR
   b. SBR to NR
   c. EPDM to NR
8. The migration rate is the fastest for 6PPD but slower for mixed diaryl phenylenediamines, and slowest for polymerized dihydrotrimethylquinoline (TMQ).
9. In the category of processing aids, oil solubility parameters are in the range of 15.0–16.5 MPa$^{1/2}$. Most processing aids have a Tg between −45°C and 90°C.

Where this has become of considerable importance is when using components in tires as reservoirs for migratory chemicals such as antiozonants and antioxidants. For example, the apex or shoulder wedge can be saturated with migratory antiozonants. As the protectants in the sidewall compound become depleted, then antiozonants from internal tire components can migrate into the sidewall, thereby offering extended protection from weathering, ozonolysis, and oxidation. This in turn contributes to an extension of the tire service life.

7.11 EFFECT OF BLENDING ON COMPOUND PHYSICAL PROPERTIES

Generally, most of the reported studies on elastomer blends focus on general purpose elastomers, with emphasis on processing and attainment of specific mechanical properties (14). It would therefore be appropriate to review these two factors. There are three methods by which a polymer blend may be mixed with carbon black.

1. Mix the first polymer with carbon black, and then dilute it with the second polymer.
2. Preblend the two polymers and then add carbon black
3. Mix a carbon black–polymer masterbatch for each of the two polymers and then blend the materials

Uneven distribution of carbon black can be observed, particularly in the first two approaches, with a potential loss in material mechanical properties. Furthermore, during processing, and when the compound is in the uncured state, carbon black can migrate from the polymer with a low affinity for carbon black to the one with a higher affinity for carbon black. This can be partially prevented by use of the concept of dynamic vulcanization where initial pre-vulcanization can lock carbon black into a particular phase. It would now be appropriate to describe the impact two specific polymer blends have on compounded material properties. Addition of BR to NR can result in:

1. Improvement in heat stability,
2. Decrease in tensile strength,
3. Decrease in 300% modulus,
4. Increase in elasticity,
5. Increase in abrasion resistance,
6. Improvement in tread wear resistance, and,
7. Improvement in fatigue resistance.

BR/NR blends tend to find application in truck tire tread compounds and sidewalls. BR is also blended with SBR in treads. Here, blend ratios tend to be optimized for factors such as wear resistance, traction requirements, rolling resistance, and damage resistance. Empirical guidelines for determining blend ratios are based on the work of Nordseik and coworkers reviewed earlier (6). Briefly:

1. An increase in SBR levels relative to BR allows improvement in traction due to the increase in compound Tg,
2. Increase in the SBR level improves handling, again due to the increase in Tg and an increase in stiffness, or storage modulus, E',
3. A decrease in SBR level with an increase in BR level improves wear performance,
4. Blends of NR, SBR, and BR can be optimized for high- performance tire treads, depending on requirements for factors, such as aligning torque, steering precision, and high speed.

7.12   SECONDARY POLYMER BLENDS SYSTEMS

NR or synthetic elastomers can be blended with a range of additional materials which can be defined as secondary polymer systems. This would include such materials as resin systems, high-molecular-weight green strength promoters, and polymeric antioxidant systems. Of these types of compounding materials, resins are the most important. Resins used in rubber compounding can be classified into one of three groups (15, 16):

(i) Tackifying resins, which remain predominantly unchanged during the vulcanization process,
(ii) Reinforcing resins, where the resin, when mixed with a catalyst or hardening agent, react with one another during vulcanization to form an interpenetrating network and do not participate in other crosslinking reactions to any significant degree,
(iii) Curing resins, where the resin reacts with the polymer, thereby crosslinking it. There are exceptions to all three definitions. For example, C5 tackifying resins can exhibit a significant level of unsaturation, which can crosslink at curing temperatures of 160°C.

7.12.1 TACKIFYING RESINS

Although there are many types of tackifying resins available, they can all be grouped into two major categories: natural resins or synthetic resins. Natural resins could include pine gum and terpenes obtained from the tapping and distillation of pine tree stumps. Coumarone-indene resins are produced by acid-catalyzed polymerization of coumarone-indene fractions obtained from coal tar, which, in turn, are a by-product of coke manufacturing. Such resins, however, are in decline due to limited availability of raw material and the ease of obtaining hydrocarbon resins from petroleum feedstocks. Synthetic resins can, in turn, be classified into two broad categories, namely C5 aliphatic resins (from C4–C6 petroleum feed streams) and C9 aromatic resins (from C8–C10 petroleum feed streams). Characteristics of tackifying resins influence not only how they will be used, but also the properties they will impart when employed within a rubber compound. These resin characteristics include:

1. Glass transition temperature,
2. Viscosity,
3. Molecular weight and molecular weight distribution,
4. Solubility and compatibility,
5. Residual unsaturation,
6. Color,
7. Acidity and basicity.

These properties will, in turn, control the performance of the resin in a specific rubber formulation and the compound processing characteristics.

7.12.2 REINFORCING RESINS

Reinforcing resins form an interpenetrating network (IPN) and can interact with other polymers. Hexamethylenetetramine–resorcinol systems do not start to react until temperatures exceed 120°C. At vulcanization temperatures of 160°C, they can significantly increase the stiffness of rubber compounds, though under dynamic service conditions such as in a tire tread compound, the stiffness imparted by such may be degraded.

Phenolic resins and high-styrene SBR resins are used for reinforcing and increasing the hardness and modulus of rubber compounds. Resorcinol resins are used as a part of the adhesion system between rubber and brass-plated steel cord or organic fibers. Both phenolic and resorcinol resins require the addition of a methylene donor such as hexamethoxy methylmelamine (HMMM) or hexamethylenetetramine (HMTA) to fully crosslink and become a thermoset. Phenol, alkyl phenols, and resorcinol can be reacted in bulk or in a polymeric formulation with methylene donors. Typical donors are 2-nitro-2-methyl propanol (NMP), HMTA, and HMMM to produce a thermoset resin network in the rubber compound. HMTA and HMMM are the preferred methylene donors used in rubber formulation.

Both of these curatives are added in the lower-temperature, final mixing stage. HMTA must be isolated from the other rubber curatives during storage and batch preparation since its basicity can cause premature decomposition of the rubber cure accelerators and can accelerate the conversion of insoluble sulfur to the soluble form. The structure of HMTA and the reaction with resorcinol are illustrated in Figure 7.8. Classical chemical studies indicate that as much as 75% of the nitrogen remains chemically bonded to the rubber, though some ammonia is released during the cure of the resin and the rubber, which can have detrimental effects on rubber composites reinforced with brass-coated steel cords.

The base structure of HMMM and the reaction with resorcinol are shown in Figure 7.9. Although it is conceivable that a methylene group might be transferred

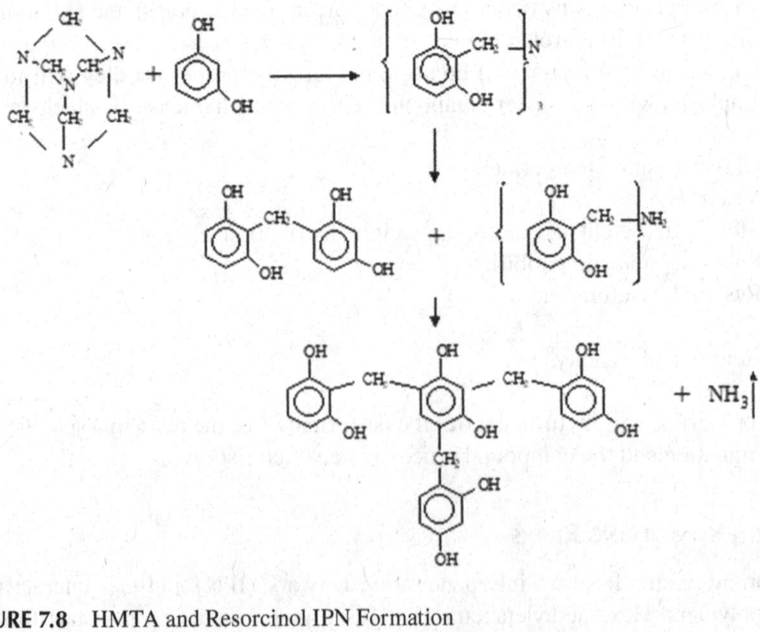

FIGURE 7.8   HMTA and Resorcinol IPN Formation

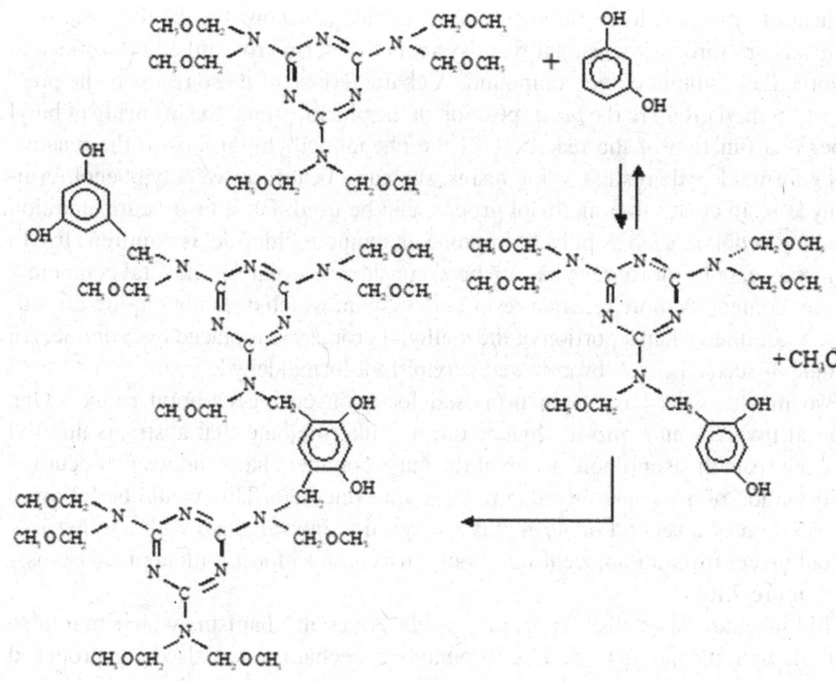

FIGURE 7.9 HMMM and Resorcinol IPN Formation

from the HMMM to the resorcinol, as is the case with HMTA, it is generally accepted that the entire melamine structure is joined to the resorcinol molecule through a methylene bridge. Formation of linear and branched structures occurs through one of two paths, either the addition of a second and third melamine methylene bridge to the same resorcinol or the reaction of additional resorcinol units onto the remaining site of the original melamine unit. Steric factors and chemical reactivities will determine the extent to which of the pathways is followed in the formation of the branched resin structure (15).

7.12.3 CURING RESINS

A significant advance in improving the service life of tire curing bladders was the development of phenol formaldehyde-based curing resins for vulcanization of butyl rubber. The resulting butyl rubber compound has a stable crosslink network and, as illustrated in this study, is not readily susceptible to reversion. Bladder compound curing temperatures of 190°C are higher than typically seen for resorcinol and HMT or HMMM systems. However, if bladders have been over-cured, they may be susceptible to tearing. Under-cured butyl bladder compounds may result in wrinkles which would then deform green tires at the start of the tire curing operation. Castor oil is typically used in bladder compounds to adjust viscosity as it is less volatile than paraffinic oils and can reduce the tendency for compound scorching.

Phenol formaldehyde curing resins are classified as resols. Unlike the linear resins, resols are three-dimensional resin systems that form crosslinked networks and reinforce the resultant cured compound. A characteristic of these resins is the presence of an alkyl group in the *para*- position of the phenol group. Resin curing of butyl rubber is a function of the reactivity of the phenolmethylol groups in the reactive phenol formaldehyde resin. Curing resins, such as a heat-reactive octylphenol formaldehyde resin containing methylol groups, can be used. For a faster cure, addition of a halogen donor, such as polychloroprene or stannous chloride, is required (16). In this instance, polychloroprene has not been considered as part of the total compound polymer content. A more reactive resin-cure system, which does not require an activator, is obtained when a portion of the methylol groups are replaced by bromine. An example of such a resin is brominated octylphenol formaldehyde resin.

Two mechanisms have been proposed for resin curing of butyl rubber. One proposal involves an *ortho*-methylene quinine intermediate that abstracts an allyl hydrogen from the isoprenoid group of the butyl polymer chain and would occur *via* the formation of a six-membered ring '-ene' intermediate. This would be followed by formation of a second *ortho*-methylene quinine intermediate with another isoprenoid group from an adjacent butyl polymer chain to form a monophenol crosslink (Figure 7.10).

This mechanism is distinct from a Diels–Alder mechanism, which may also occur, though, if so, in low yield. A chromanone mechanism has also been proposed and leaves no unsaturation upon crosslink formation. Though this latter mechanism has now become the most likely one, the six-membered '-ene' reaction mechanism may still play a role in crosslink formation but, again, at a low yield.

Figure 7.11 is a schematic illustrating the reaction of a curing resin (heat-reactive octylphenol formaldehyde) in the vulcanization of butyl rubber. Following the elimination of water in the reaction sequence, the exomethylene group and carbonyl oxygen react with an isoprenyl unit in butyl rubber to form a chromanone ring. Chromanone ring structures are very stable and are frequently found in natural product biosynthesis.

Given the current understanding of the mechanism of the resin vulcanization, the purpose of this study has been to focus on the technological aspects of resin-cured butyl rubber compounds, such as those used in curing bladders. This has involved investigating the effect of polymer viscosity (i.e., molecular weight) and isoprene unsaturation level on the curing performance and mechanical properties of cured compounds. These two polymer properties are the primary variables that will govern the final properties of the compound. Butyl 268 is also the preferred polymer for curing bladders.

7.12.4 POLYMERIC GREEN STRENGTH PROMOTER

A number of polymeric green strength promoters have been developed. The green strength promoter with commercial designation, Vestenamer 8012, is a polyoctenamer manufactured by Evonik GmbH of Germany. The raw material for this polymer is cyclooctene. In a metathesis polymerization reaction, cyclooctene is converted into a polyoctenamer. It was developed as a green strength promoter for tire sidewalls,

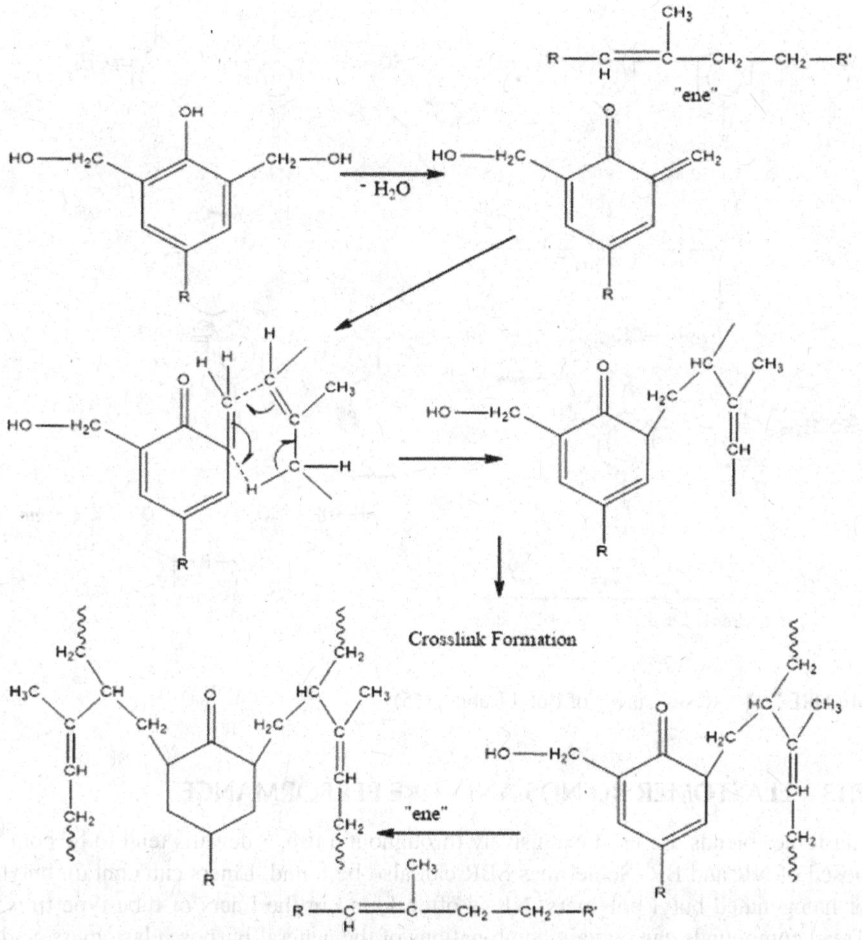

FIGURE 7.10 Allyl Hydrogen Mechanism (15)

apexes or bead fillers, and other components where dimensional stability is important. It meets four requirements: serving as a green strength promoter; rapid crystallization rate, low viscosity, and unsaturation for participation in the vulcanization or crosslinking reaction.

Formation of a network of low-molecular-weight polymers will cause a loss in mechanical properties. However, a network of unsaturated, cyclic, low-molecular-weight polymers will have the potential of improving mechanical properties of the material in a composite structure, such as a tire. This is due in part to participation in vulcanization or chemical cross-linking. The high macrocyclic ring structure of polyoctenamers improves heat stability at elevated temperatures, due to the interpenetration of linear polymer chains and cyclic macromolecules. In conclusion, a more in-depth review of casing and inner liner compound polymer systems can be found in references (17) and (18).

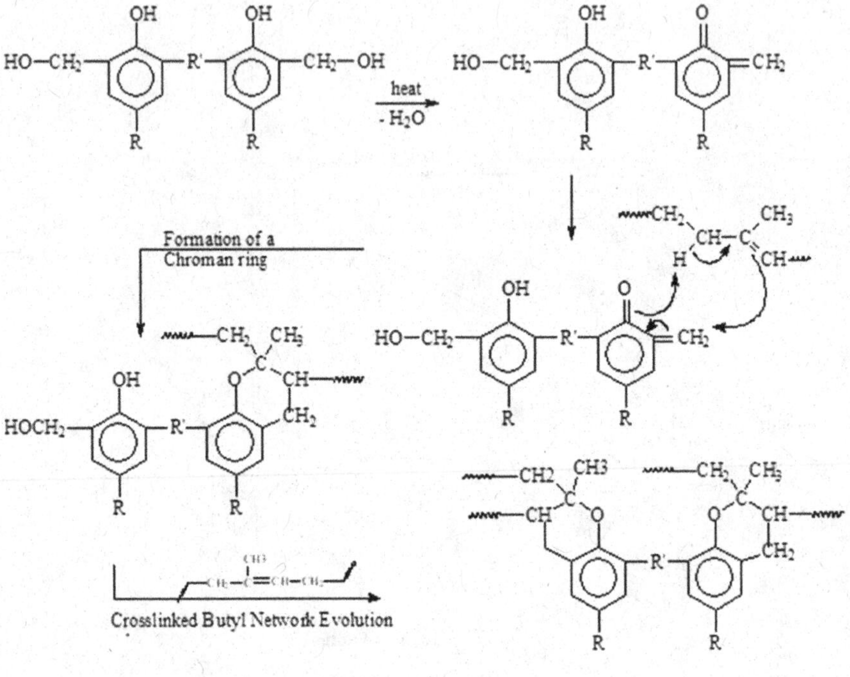

FIGURE 7.11   Resin Curing of Butyl Rubber (15)

7.13   ELASTOMER BLENDS AND TIRE PERFORMANCE

Elastomer blends are used extensively throughout a tire. Sidewalls tend to be composed of NR and BR. Sometimes SBR can also be found. Liners can contain butyl or halogenated butyl polymers. NR is often found in the liners of tube-type tires. Tread compounds can contain combinations of the general-purpose elastomers, and additional specialty polymers, such as the ter-polymer, styrene-isoprene-butadiene rubber, or SIBR. Within one class of polymer such as BR, the microstructure can have a significant impact on performance. For example, BR can be polymerized by a range of metals such as lithium in solution polymerized systems, and nickel or titanium in Ziegler-Natta systems. Here, the *cis*-butadiene level can range from 43% to 98%. As an example of how this affects compounded properties, fatigue life of a BR in a carbon black-loaded compound can vary from 3 kilocycles to failure on the Monsanto Fatigue to Failure instrument to over 40 kilocycles to failure. Furthermore, an increase in the vinyl-butadiene level will tend to have a negative effect on resistance to fatigue. This in turn impacts tire performance. As a general rule, several guidelines can be outlined to predict abrasion resistance of a tread compound:

1. 100% BR gives the best tread wear performance or abrasion resistance.
2. SBR tends to give directionally better wear performance compared to NR.
3. Blending NR or SBR with BR gives an improvement in abrasion resistance or tread wear.

4. Similarly, the tread wear performance of NR can be improved through addition of SBR.
5. A balanced set of tire tread material properties, such as wear, traction, heat build-up, and damage resistance, is achievable through effective blending of all three general purpose rubbers, namely NR, SBR, and BR.
6. The properties of compounded blends will be affected by the vulcanization system and filler distribution
7. Attainment of a single Tg is important in ensuring low hysteresis.

Though preparation of blends can enhance material performance properties, there are losses in other characteristics. Addition of BR to NR compounds results in a decrease in tensile strength, and in ultimate elongation and tear strength. For example, addition of 50 phr of BR to an all-NR compound, and having equal carbon black dispersion between the two polymers, tensile strength decreases from 24 MPa to 17.5 MPa. With all the carbon black in the natural rubber phase only, the tensile strength drops further to 15.8 MPa (Table 7.6). Similar trends are observed in compounds with only 25 phr of BR added to the NR (Table 7.7). Tensile strength drops from 23 MPa to 18 MPa, with the location of the carbon black playing a critical role in generation of the compound mechanical properties. Several points can be noted here:

1. Increase in the amount of carbon black in the BR phase increases the compound tensile strength, tear strength, and resilience.
2. Adding all the carbon black to the NR phase causes a drop in tensile strength of up to 25%
3. Hardness increases with more of the carbon black in the NR phase.

TABLE 7.6

Natural Rubber Polybutadiene Blends and Effect of Carbon Black Location on Compound Mechanical Properties (1, 4, 19)

Compound		1	2	3
Natural rubber		100.0	50.0	50.0
Polybutadiene			50.0	50.0
Carbon black N60		40.0	40.0	40.0
Carbon black location		NR	BR	NR
Tensile strength	MPa	24.5	17.5	15.8
Elongation at break	%	511.0	435.0	439.0
300% Modulus	MPa	12.2	10.8	9.3
Tear strength	kN/m	18.7	14.8	11.3

Note: Tensile strength measured according to ASTM D412, Tear strength to ASTM D624.

TABLE 7.7

Natural Rubber Polybutadiene Blends (1, 4)

Compound		1	2	3
Natural rubber		100.0	50.0	50.0
Polybutadiene			50.0	50.0
Carbon black N60		40.0	40.0	40.0
Carbon black in NR	%	100.0	25.0	75.0
Carbon black in BR	%		75.0	25.0
Tensile strength	MPa	23.1	19.1	17.4
Elongation at break	%	509.0	451.0	444.0
300% Modulus	MPa	11.5	10.7	10.1
Tear strength	kN/m	26.5	19.7	18.4

In summary, several points can be noted in preparation of elastomer blends. Briefly:

1. Proper selection of polymers to be used in elastomer blends is critical.
2. The impact of microstructure, as determined by the polymer catalyst system, will influence the selection of polymers.
3. Distribution of compound ingredients, such as carbon black, affects key mechanical properties, such as tensile strength.
4. Distribution of cure system components is affected by the selection of polymers in the blend.
5. High BR levels favor good tread wear performance, but there will be a trade-off with other properties, such as traction.

7.14 SUMMARY

This review has briefly discussed aspects of basic polymer science, tire compounding, and blending of elastomers. The impact of elastomer blends on tire performance and secondary polymer blend systems was also discussed. Phase morphology is influenced by the polymers used in a blend. Compounding of elastomers will have a significant effect on tire performance. Figure 7.12 collates the observations presented in this review into one diagram. For example, if rolling resistance and tread wear is preferred, then the polymers selected will tend to be low-styrene, low-vinyl SBR, NR, and high *cis*-polybutadiene. Conversely, if a high-performance tire tread compound is required, then there would be a predominant use of high-styrene SBR with high-vinyl BR, depending on the degree of traction required. Building on this illustration, Figure 7.13 shows a guideline for heavy-duty tire tread compound polymer selection, emphasizing addition of NR for better tear strength and, in turn, better tire tread cutting and tearing resistance.

Application of novel approaches to raw material development and their use in elastomer blends could offer potential for the next breakthrough in polymer science.

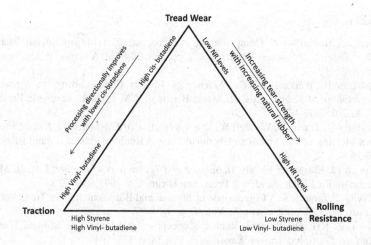

FIGURE 7.12 Performance Triangle and Elastomer Selection for Passenger Tire (PCR, RLT) Tread Compounds

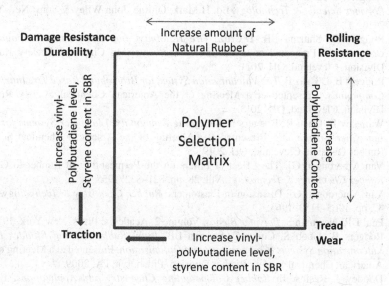

FIGURE 7.13 Performance Diagram and Elastomer Selection for Truck Tire (TBR, OTR) Tread Compounds

To conclude, a wide range of subjects in the field of elastomer blends offer opportunities for further research. These include:

1. Block copolymers.
2. Polyurethane and general-purpose elastomer blends.
3. Thermoplastic-elastomer blends, and finally,
4. Application of novel monomers, such as organosilanes, and organometallic complexes, such as those of zirconium, titanium, and cobalt.

REFERENCES

1. Rodgers B, Halasa A. Compounding and Processing of Rubber/Rubber Blends. In *Encyclopedia of Polymer Blends*. Vol. 2, pp. 163–204. Ed. A Isayev. Wiley, New York. 2011.
2. Rodgers B, Waddell W. The Science of Rubber Compounding. In *Science and Technology of Rubber*. Eds. JE Mark, B Erman, CM Roland. Academic Press, New York. 2015, 417–470.
3. Rodgers B, Tracey D, Waddell W. *Tire Application of Elastomers, 1 Treads*. Presented at a Meeting of the American Chemical Society Rubber Division, Grand Rapids, MI. 2004.
4. Corish PJ. Elastomer Blends. In *Science and Technology of Rubber*. Eds. JE Mark, B Eurman, FR Eirich. Academic Press, San Diego, CA. 1994, 489–525.
5. Day G, Futamura S. A Comparison of Styrene and Butadiene in Tire Tread Polymers, *Kautschuk Gummi Kunststoffe*, Vol. 40, p. 39. 1987.
6. Nordsiek KH. The 'Integral Rubber' Concept – An Approach to an Ideal Tire Tread Rubber. *Kautschuk Gummi Kunststoffe*, Vol. 38, p. 178. 1985.
7. Halasa A. Preparation and Characterization of Solution SIBR via Anionic Polymerization. *Rubber Chemistry and Technology*, Vol. 70, p. 295. 1997.
8. Rodgers B, Tallury, S, & Klingensmith W. Rubber Compounding. In *Encyclopedia of Polymer Science & Technology*. Ed. H Mark. Online. John Wiley & Sons, New York. 2016.
9. Rodgers B, Sharma BB, D'Cruz. *Tire Halobutyl Innerliners and Compounding Components*. Presented at a Meeting of the American Chemical Society Rubber Division, Cleveland, OH. 2011.
10. Rodgers B, D'Cruz B. *The Vulcanization System for Bromobutyl Based Tire Innerliner Compounds*. Presented at a Meeting of the American Chemical Society Rubber Division, Cleveland, OH. 2013.
11. Wamsley E, Chen R, Rodgers B. *Designing Bromobutyl Elastomer Systems for Tire Innerliner Compounds*. Presented at a Meeting of the American Chemical Society Rubber Division, Cleveland, OH. 2019.
12. Van Amerongen GJ. The Effect of Fillers on the Permeability of Rubber to Gases. *Rubber Chemistry & Technology*, Vol. 28, pp. 821–832. 1955.
13. Van Amerongen GJ. Diffusion in Elastomers. *Rubber Chemistry & Technology*. Vol. 37, pp. 1065–1152. 1964.
14. Paul DR, Newman S. *Polymer Blends*. Volume 2. Academic Press, New York. 1978.
15. Rodgers B, Jacob S, Curry C, Sharma BB. *Butyl Rubber Curing Bladder Resin Vulcanization Systems: Compositions and Optimization*. Presented at a Meeting of the American Chemical Society Rubber Division, Pittsburgh, PA. 2009.
16. Duddey J. Resins. In *Rubber Compounding Chemistry and Applications*. Ed. B Rodgers. CRC Press/Taylor & Francis, Boca Raton, FL. 2015, 379–418.
17. Waddell W, Rodgers B. *Tire Applications of Elastomers; Casing*. Presented at a Meeting of the American Chemical Society Rubber Division, Grand Rapids, MI. 2004.
18. Waddell W, Rodgers B. *Tire Applications of Elastomers; Innerliner*. Presented at a Meeting of the American Chemical Society Rubber Division, Grand Rapids, MI. 2004.
19. Rodgers B, Ryba J. *Principles of Polymer Blending*. Presented at the 147th Meeting of the Rubber Division, American Chemical Society, Philadelphia, PA. 1995.

8 Tire Manufacturing

8.1 INTRODUCTION

Consistent with the underlying complexity of the modern pneumatic tire, manufacturing operations are similarly elaborate, with complex operations requiring highly skilled work forces. Despite increasing levels of automation, tire production is still very labor intensive. The most efficient tire factories are typically divided into six departments or unit operations that perform specific tasks (Figure 8.1). In addition, a secondary set of support operations would include the laboratory, engineering and quality control, utilities, and administration. Global tire manufacturers have different manufacturing philosophies in that they may set up independent, self-sufficient, operations on a single site, or cluster the factory operations locally to serve a region, e.g., a separate mixing facility supporting several tire production plants. They may also use only one specific type of compound-mixing system or tire-building machine, or factories may be configured to produce only one type of tire, such as heavy-duty truck tires.

However, in considering what is common to all manufacturing plants, factories are typically designed to build a minimum of 25,000 passenger tires (PCR) per day, 270 tons of finished product, or 4,000 truck tires (TBR, truck and bus radials), or their equivalent. This is the point where manufacturing efficiencies tend to become more favorable (Figure 8.2). At 70,000 PCR units, or equivalent, cost efficiencies have been fully captured and manufacturing complexities then tend to negate any further benefit resulting from plant size increases.

8.2 RAW MATERIALS RECEIVING AND COMPOUND MIXING

Raw materials are received at the tire factory under a purchasing specification. This specification is typically prepared by the company's central Technology team and almost universally consists of four sections:

1. Introduction, with any coded identifiers, type of material, and the chemical name
2. Description, including color, form, and compliance test data that the supplier has committed to meeting, such as trace metal content and color
3. Material acceptance requirements and associated specification test data validated by the supplier, such as MDR rheometer or Mooney viscosity data, and target test data that the receiving plant will verify. In addition, there may be additional baseline test specifications, such as compound tensile strength and other specialty testing
4. Shipment compliance and contract obligations, including packaging requirements, reference formulations, and controls over changes in manufacturing processes and sites.

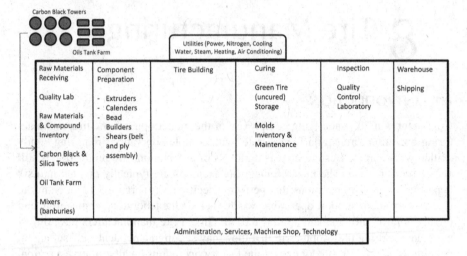

FIGURE 8.1   Tire Factory Unit Operations (1)

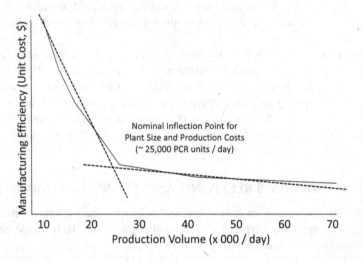

FIGURE 8.2   Nominal Cost-Reduction Profile as Tire Plant Capacity Increases (2) (Each tire company has a different cost structure but would follow a similar capacity vs cost profile)

Most tire companies will code their raw materials for a variety of reasons, such as simplification, to facilitate programming of computerized factory compound formulation and mixing, for security of formulation compositions, and to provide a common link to the specific purchase specification and other quality control documents.

Raw material testing is also one of the three compound production quality control test points, namely the individual raw material, the final mixed "green" compound, and the final cured compound, the properties of which are to be checked before being used in tires. The specific test protocols will differ for each manufacturer but will be designed to address their own quality assurance requirements and to ensure that the specific tires meet the design requirements and perform to the market needs.

Mixing of a compound is the process of blending together all of the materials in a formulation to produce a homogeneous material. Internal mixers used to conduct the operation consist of two counter-rotating rotors in a large chamber, that apply a high degree of shear, at temperatures up to 180°C, to the raw materials (Figure 8.3, Table 8.1). The large tire manufacturing companies tend to standardize on one type

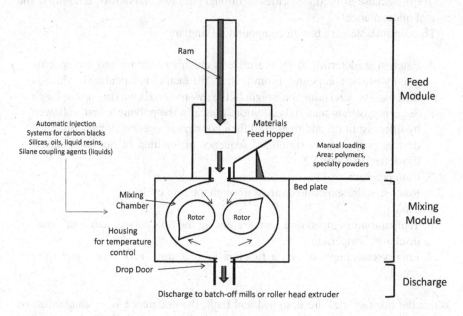

FIGURE 8.3 Schematic of an Internal Mixer or Banbury (3)

TABLE 8.1
Internal Mixer Descriptions Used in Tire Manufacturing (2, 4)

Machine Designation	Application	Chamber Volume	Nominal Batch Production Weights		Rotor Speed	Ram Pressures
		L	kg	lb	RPM	
Midget	Laboratory	0.4	0.3	0.7		
BR	Laboratory	1.6	1.1	2.6		
OOC	Laboratory	4.3	3.2	7.0		
F3D	Specialty compounds	70.0	53.0	116.0	50–100	60–80 psi
F80	Specialty compounds	80.0	60.0	132.0	35–105	60–80 psi
9	Specialty compounds	186.0	145.0	316.0		60–80 psi
11D	High-volume production	234.0	177.0	390.0	30–60	60–80 psi
F270	High-volume production	270.0	202.0	445.0	20–80	60–80 psi
F370	High-volume production	370.0	277.0	600.0	20–60	60–80 psi
F620	High-volume production	620.0	465.0	930.0	30–60	60–80 psi

of mixer from a global supplier, such as Kobelco, which is part of the Kobe Steel Group in Japan, or Harburg-Freudenberger Machinery (HF Group) from Germany and the USA. Table 8.1 illustrates some of the general capabilities of Farrel-Bridge Banbury mixers that are typically found in tire production plants. Globally, F270-sized machines would be the most popular in tire manufacturing plants, followed by the F370, because of their capacities or throughput rates, versatility, reliability, and ease of maintenance.

The controllable variables in compound mixing are:

1. Batch size, determined by the mixer's chamber volume and the specific gravity of the compound. From this, a "fill factor" is calculated, which is then used to determine the weight of the raw materials making up the batch
2. Sequence of raw materials addition, typically the polymers first, followed by fillers, split oil additions, and then powders in pre-weigh bags (for anti-dusting purposes). A simplified sequence of addition of raw materials is illustrated in Table 8.2
3. Ram pressure
4. Rotor speeds, sometimes varied through the mix cycle
5. Time
6. Temperature (temperature increase during the mix period, and final batch discharge temperature)
7. Energy consumption, i.e., a function of pressures, mix times, and rotor speed.

When the tire factories are designed and built, the preference is to standardize on one type of mixer, where a necessary variable is the rotor type. There are two types of rotors found in internal mixers, tangential or intermeshing (Figure 8.4). For silica-loaded compounds, much higher shear is required to obtain adequate filler dispersion and compound homogeneity. In such cases, mixers with intermeshing rotors are preferred. In some instances, there may also be a need to dedicate specific mixing lines for compounds, such as the inner liner, to avoid any cross-contamination.

In many high-volume mixing operations, batches are discharged when the mix cycle reaches a target temperature. For carbon black-filled compounds, this can be up to 180°C, though 160°C to 170°C might be more typical. In the case of new-technology silica-filled compounds, batch discharge temperatures should not exceed

TABLE 8.2
Mixing Stages in a Compound Mix Cycle

Stage	Type	Materials Typically Added by Stage
1	1st Non-productive	Polymer, peptizer
2	2nd Non-productive	Carbon black, oils, waxes, resins, etc.
3	3rd Non-productive	Antioxidants, antiozonants
4	4th Non-productive	Re-mill, if needed (drop at 120°C)
5	Productive	Vulcanization system (drop at 105°C)

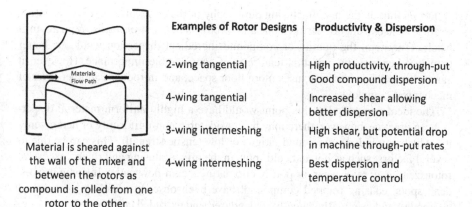

	Examples of Rotor Designs	Productivity & Dispersion
	2-wing tangential	High productivity, through-put Good compound dispersion
	4-wing tangential	Increased shear allowing better dispersion
	3-wing intermeshing	High shear, but potential drop in machine through-put rates
	4-wing intermeshing	Best dispersion and temperature control

Material is sheared against the wall of the mixer and between the rotors as compound is rolled from one rotor to the other

FIGURE 8.4 Rotors used in Internal Mixers at Tire Plants

160°C, and, at the final stage, when the vulcanization system components are added, the discharge temperature would typically not exceed 105°C.

Elaborating, compound mixing typically consists of between two and four stages, which is required to incorporate the raw materials in the compound. The shearing action can generate considerable heat, so both rotors and chamber walls have both steam heating and water-cooling capability to maintain a controlled temperature high enough for blending and mixing but low enough to prevent the premature onset of vulcanization, which is sometimes referred to as scorch.

Some compounds such as bromobutyl inner liners will have a two-stage mix cycle, typically referred to as the first masterbatch or non-productive, which is discharged at 125°C, followed by a final, productive stage, where the vulcanization system materials, i.e., sulfur and accelerators, are added and have a batch discharge temperature of between 90 and 105°C, similar to other compounds.

For tread compounds, a four-stage mix cycle is more typical, with three non-productive stages and a final productive stage. In some instances, with more complex formulations, such as high-performance silica tread compounds, up to seven stages can be involved, including "re-mills", where the compound is passed through the mixer at typically 120°C for 90 to 120 seconds to optimize compound viscosity.

After mixing, the rubber charge is dropped into an extruder feed box underneath the mixer and fed by the extruding screw into a roller die. Alternatively, the batch can be dropped onto an open rubber mill batch-off system. A mill consists of twin counter-rotating rolls nominally 50 cm in diameter and 2 m in length. One roll may be serrated to provide additional mechanical shearing to the rubber, thereby further improving dispersion of the compounding ingredients. A compounded rubber sheet will then be produced, although the sheet can also be taken off the mill in the form of a strip. The sheet or strip is cooled in a water tank, normally dusted with an anti-tack agent, such as calcium carbonate, zinc stearate in some instances, talc or soft clay, passed through a cooling and drying festoon, and then passed to a lay-down on a pallet at a target temperature below 40°C.

Depending on the manufacturing philosophy of the company, there can be either a roller- head extruder or between one and three sets of batch-off mills. With a 3-mill batch-off system, the quality of compound ingredient dispersion would be better than in other systems, and this feature is of considerable importance. However, it can be more labor intensive, uses more floor space, and introduces another layer of mechanical complexity.

The ideal compound at this point would have a highly uniform material dispersion. In practice, however, there may be considerable non-uniformity or heterogeneity, which can induce compound fatigue or low tensile strength. This can be due to several factors, such as hot and cold spots in the mixer housing and rotors, excessive rotor clearance, rotor wear, or poorly circulating steam flow paths. In some mixers, dead spots, causing trapped compound, have been observed, leading to scorched compound and eventually defective calendared and extruded components.

The current trend in new manufacturing plants is to replace the batch-off mill with a roller- head extruder, thereby reducing labor requirements and improving throughput rates. In an automated plant, where a roller-head extruder replaces the batch-off mill systems, other than one operator or inspector for 7 to 10 mixing-line extruders, there would be no labor requirements until the compound reaches the lay-down stage.

A compound mixing line design in a large tire factory can extend for up to 100 m (Figure 8.5). A tire factory producing 25,000 tires per day or over 200 tons of finished compound would typically have five mixing lines, with one potentially dedicated to mixing halobutyl inner liner compounds.

In addition, the mixing operations will also consist of:

1. Raw material inventory warehousing, preferably fully automated.

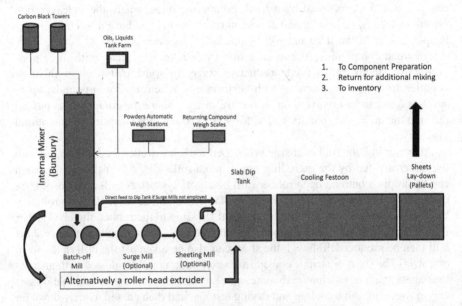

FIGURE 8.5 Flow Path Schematic of Mixing Line (Mill Batch-off System)

2. Hot houses, typically three rooms, of which two are used for pre-heating natural rubber before mixing and one for halobutyl. Temperatures are typically held at around 55°C. In all cases, strict control of hot room residence times is of the utmost importance.

3. For bromobutyl compounds, hot houses also assist in better pigment/oil incorporation. Hot house temperature must be less than 50°C as prolonged exposure of BIIR polymers to this temperature will lead to temperature-induced Mooney creep.

4. Automatic weigh-up systems for powders along with the silos for carbon blacks and silicas, and a tank farm for oils, resins, and other liquids, such as silane coupling agents.

5. Mixing lines and final compound inventory area.

6. Quality Control laboratory.

7. Mixing area Operations Control Center.

In many instances, and before being used in a compound formulation, natural rubber may be pre-masticated. The rubber or blends of different natural rubber grades, depending on the application, would be added to the mixer with a peptizer. The polymers are mixed and then sheeted as a new raw material code. This results in a more consistent polymer with a more uniform viscosity, in turn, allowing mixing efficiencies when compounded with carbon black, silica, and the remaining compounding ingredients.

Mixing is one of the most important operations in tire manufacturing. A number of precautions and troubleshooting guidelines will be categorized into one of four topics: maximization of dispersion and homogenization, prevention of scorched compound, avoidance of contamination, and cross-blending or disposition of scrap. In addition, the engineer must ensure consistency and batch-to-batch uniformity. Potential problems include:

1. Inadequate dispersion; multiple causes including
   a. Insufficient mixing time, inappropriate sequence of raw material addition, or incorrect batch size (fill factor)
   b. Wrong rotor speed, ram pressure, or temperatures
   c. Excessive rotor or chamber wear. An internal mixer would require inspection for refurbishment every seven to eight years, or, in the case of silica mixing, every five years
   d. Moisture, oil addition point
   e. Absence of a homogenizing agent, such as Struktol 40MS™, which is a very effective dispersion aid

2. Scorched compound; could be due to
   a. Excessive rotor speeds and high temperature
   b. Batch set-up, i.e., inappropriate sequence of addition of raw materials
   c. High final batch lay-down temperatures which should not exceed 40°C

3. Contamination:
   a. Industrial hygiene concerns
   b. Inadequate clean-outs during compound mix changeovers

    c. Mixer rotor seal leakage
    d. Raw material addition errors
4. Batch-to-batch variability
    a. Variability in milling time, temperatures, or energy input
    b. Variation in work-away levels or cross-blending when using intermediate surge mills.

8.3 COMPONENT PREPARATION

Tire component preparation falls into one of three general operations, namely extrusion, calendaring, and bead building. Typically, this area in the factory, where the tire components are produced, is both humidity and temperature controlled (ambient or nominally 25°C).

The number of extruder lines in a factory depends on the size and product complexity, but a typical factory producing 25,000 PCR tires per day would have three tread extruder lines and two sidewall extruders. The sidewall lines will be duplex or triplex units to co-extrude the shoulder wedge or other components, depending on the tire complexity and the factory production portfolio. Similarly, the tread extruder lines will be triplex or higher, i.e., quadruplex (four barrels) or pentaplex (five barrels). In addition, there may be some additional, smaller extruder lines to produce apexes and shoulder wedges as separate components.

There are two classes of extruders in tire plants, hot feed and cold feed. Hot-feed extrusion lines, found in older plants, are fed by a series of two-roll mills. The compound is first added to a "break-down" mill, where it is warmed, masticated, and then fed to a second mill *via* an overhead conveying system. Compound is then fed as a uniform strip to the extruder, or, alternatively, to another set of two blending mills before going to the extruder. Mills are thus typically arranged in pairs. In some older plants, as many as ten mills or five pairs of mills have been configured to supply compound to a large tread extruder line. Such lines would be found where high-viscosity natural rubber tread compound extrusions were being produced. The extruder in such lines tends to have short barrels with a length to diameter (L/D) ratio as low as 4:1 (Figure 8.6).

In newer factories, cold-feed extruders with multiple barrels are found. Cold-feed extruders have much longer barrels, due to the need to increase the temperature of the compound, thereby lowering compound viscosity and enabling better flow shaping of the final profile. L/D ratios can be as high as 24:1. Cold-feed extruders will also have vents to expel low- molecular-weight volatile organic compounds which, in the case of hot-feed extrusion lines, would have occurred at the break-down mills and when hot compound is fed into the extruder feed box. Reasons for the shift to cold-feed systems include i) increasing tread profile complexity, ii) lower labor requirements, iii) lower floor space requirement, iv) lower volatile organic compound (VOC) emissions, and v) lower installation costs. A typical cold- feed tread extrusion line in a newer tire plant would consist of:

1. Compound staging area
2. Triplex extruder (three barrels for the tread compound, base compound, and mini-wing sidewall compound; Figure 8.7)

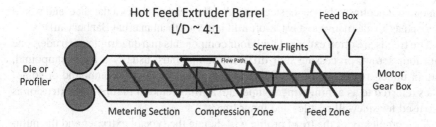

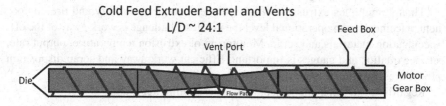

FIGURE 8.6 Extruder Barrel and Screw Configurations

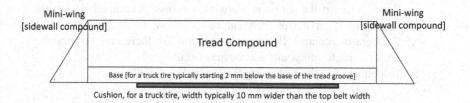

FIGURE 8.7 Triplex Tread Extrusion

3. Cushion applicator (small mill applying a hot strip, typically similar to the belt compound formulation) under the tread base before the extruded tread enters the cooling line
4. Cooling line (water spray followed by air), typically in a 3-deck configuration
5. Lay-down or tread booking station, either onto a roll or a leaf-trailer with two treads per leaf

The extruder consists of a screw and barrel, screw drive motor and gear box, heaters, and a die. Along the extruder screw there are three basic zones: i) a feed box and feed zone, where the rubber compound enters the extruder barrel, ii) a compression zone where the screw diameter increases, thereby applying pressure to the compound as it is forced through the barrel by the screw, and iii) a metering zone, where the compound under pressure enters the die and is shaped into the dimensions of the required profile. The extruder screw also provides for additional mixing of the compound through the shearing action of the screw.

The extruder die can be described by its function. The primary purpose is to form the extrusion profile, such as a tread, sidewall, or shoulder wedge. In addition, there are two forms of die producing sheets. A roller die will form a profile, such as would be used for contoured inner liners, barrier or sidewalls. A roller head produces a flat sheet and would be found as an alternative to a batch-off mill underneath an internal Banbury mixer.

Tire treads are often extruded with four components in a quadruplex extruder, one with four screws processing four different compounds, usually a base compound, core or under-tread compound, top tread compound, and wing compound or sometimes referred to as a mini-wing compound, as illustrated in Figure 8.8. Extrusion is also used for sidewall profiles and inner liners.

The complexity of the tread profile will dictate the type of extruder and the number of barrels on the extruder. Some examples of the profiles that can be produced by a triplex, quadraplex, or pentaplex line are illustrated in Figure 8.8.

Though multiplex extrusions offer improvements in uniformity and tire component placement, at an operational level the greatest challenge is work-away of the off-specification materials and scrap. Monitoring of extrusion temperature, output rate, surface quality, and gauges is important. Efficient work-away and scrap disposition procedures are equally important.

Regarding extrusion quality, some troubleshooting items are:

1. Low output rate: Variable or low output rates can be due to two possible causes
   a. Screw wear: if the screw is worn with excess clearance between the screw flights and barrel, there can be back-flow
   b. Temperature control: if temperature is too low there may be slippage, or, if too high, compound scorch may occur
   c. Inadequate feed rate (starving the feedbox)

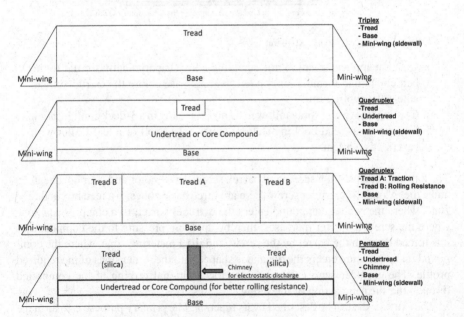

FIGURE 8.8 Compound Configurations for Three Types of Extruded Profiles (Triple, Quadraplex, Pentaplex Extruders)

2. Profile dimensional control: in addition to compound formulation, gauge and profile variation could be caused by extrusion take-away speed, screw speed variation, feed rate (starving the feedbox), and barrel temperature profile and control
3. Rough extrudates: can be due to nonhomogeneity in the compound, too-high viscosity, or compound scorch. Scorch can be caused by compound slippage due to excessive screw wear
4. Porosity: can have two causes,
   a. Air being sucked in at the feedbox due to insufficient compound, i.e., starving the feedbox
   b. Volatile raw materials in the compound formulation. For example, ethanol emissions in compounds containing silane coupling agents can create porosity if temperatures in the barrel get too high.

In component preparation there are three types of calenders,

1. Gum calenders, which are used to make rubber sheets such as inner liners, barrier or other gum strips
2. Fabric-coating lines, where the rubber compound is spread on the fabric
3. Steel cord coating for belt and truck tire ply production

The calender is a heavy-duty machine equipped with three or more chrome-plated steel rolls that revolve in opposite directions (Figure 8.9). They can have many different designs and roller configurations, such as:

1. Vertical 3-roll machine (Figure 8.9)
2. 3-roll center roll off-set vertical configuration
3. Vertical 4-roll machine (Figure 8.10)
4. Inverted "L" 4-roll machine
5. 4-roll "Z" configuration
6. Dual or duplex vertical 3-roll configuration (used for plying up two different compounds)

Calender rolls are heated with steam or circulating water; the gearing allows the rollers to operate at variable speeds like the mill rolls. Temperatures can therefore be

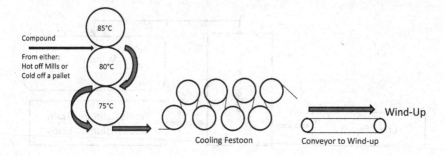

FIGURE 8.9 Vertical 3-Roll Gum Calendar Line Schematic

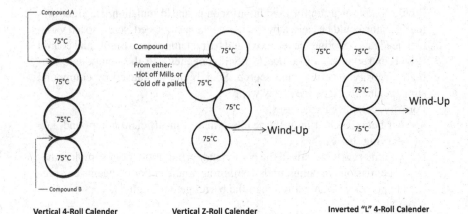

| Vertical 4-Roll Calender | Vertical Z-Roll Calender | Inverted "L" 4-Roll Calender |

FIGURE 8.10 Examples of Calender Configurations

set to obtain the best processing conditions for the various types of compounds in a tire plant. ExxonMobil has published a manual on inner liner compound production and processing, which discusses the various calendar configurations; further reference to this manual is recommended (5). Though intended as a guide in inner liner compound processing, many of the principles apply to all compounds and components in tires and so it is a valuable reference guide for tire engineers in research and development, and in manufacturing.

In fabric calendering, the fabric is passed through the calender rolls, and compound is applied above and below to fully cover the cords. The amount of compound deposited onto a fabric or steel cord is determined by the distance between the rollers and can be monitored by various types of sensor, such that each cord is coated on all sides with rubber compound.

In the case of wire calendering, the wire is first loaded onto spools inside a creel room which has tight temperature and humidity controls. Wire is then passed through a comb which sets the final wire spacing before it is coated with rubber at the calendar (Figure 8.11).

For inner liner production, the trend is to extrude, rather than calender, inner liners, due to the high cost of calender trains, with the barrier being best applied at the end of the extrusion line and before wind-up.

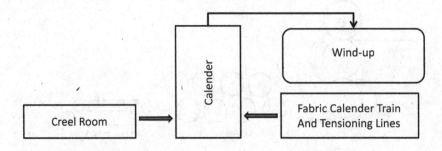

FIGURE 8.11 Schematic of Wire Calendaring

Roller die extruders and calenders have contoured rolls to ensure that the final tire inner liner gauge is uniform. Heavy centerline and shoulder gauges of the inner liner sheet allow for more uniform gauge control in the final built tire after it has gone through the casing shaping turn-up. Friction ratios, or the ratio of rotating speeds of the front and rear rolls on two-roll mills, for halobutyl compounds are the reverse of those for general-purpose elastomers, e.g., natural rubber-based compound tie gum strips. Temperatures of the rolls on both extruders and calendars are similar, typically 75°C to 85°C. Output rates depend on the compound but are of the order of 6 to 7 m/min. This does not take into consideration control of viscosity and green strength for splice integrity. Caution is also needed in controlling feed rates of extruder and banks on mills to prevent air being drawn into the compound, to subsequently form blisters

Upon completion of calendered or extruded sheets, such as liner, barriers or other gum strips, the component is plied up or wrapped with a backing material, such as a polyethylene or polypropylene sheet. This helps maintain a fresh component surface, with no loss of tack. In some instances, fabric-backing material, such as cotton and, to a lesser degree, polyester or nylon could cause a loss in compound tack, which may be desirable.

Problems encountered in calendering operations would fall into one of four areas, namely scorched compound, blisters, uneven gauge, and tack. Briefly:

1. Scorched compound, due to
   a. Inadequate temperature control, high speeds causing high shear and heat build-up
   b. Excessive compound banks either on the calendar or on feed mills, leading to high dwell times at elevated temperatures
2. Blisters
   a. Roll banks on feed mills, trapping air
3. Holes or rough uneven gauges
   a. Low compound sheet green strength
   b. Cold compound
   c. Compound temperature variability
   d. Low compound viscosity
   e. Shrinkage, due to mix cycle set-up, high polymer molecular weight, or inadequate stress relaxation before sheet wind-up
4. Tack
   a. High temperature, causing ingredient migration
   b. Low solubility of certain compounding ingredients

8.4   ELECTRON BEAM RADIATION PROCESSING

The barrier components are, in many instances, passed through an electron beam radiation (EBR) unit before going to the extrusion line for pre-assembly or directly to the building machine. Treating calendered or extruded rubber sheets, such as the barrier, increases green strength, enabling a reduction in component gauge without loss of other properties, such as tack or adhesion. In the case of the tire barrier, it is typically only irradiated on the ply side of the calendared or extruded sheet.

Payback periods or return on capital investments for EBR units are of the order of months, as a result of increasing green strength, allowing reduced component gauges. It therefore merits discussion.

There are three basic designs of electron beam radiation units: i) scanning beam units, ii) single linear filament units, and iii) multiple filament units. In all three configurations, electron beam radiation is generated in a vacuum, and emitted electrons are passed through an accelerator containing an electrostatic field, with the energy gain measured in electron volts, eV. The parameters around which electron beam processing is conducted are as follows:

1. The accelerator voltage, which is the potential difference between the anode and cathode in the accelerator, is expressed in kV or MV. EBR equipment is designed to supply voltages from 80 kV up to 10 MV. Equipment running at 250 to 750 eV may be considered typical for barrier compounds
2. The electron beam current is measured in mA
3. The electron beam power or dosage in kGy/second

The compound treatment is a function of line speed, electron beam penetration depth into the compound, and dosage rate. In irradiated rubber compounds, the polymer can be either crosslinked or degraded (Table 8.3). The amount of radiation to which the rubber is subjected is a function of the acceleration current and the depth of penetration of the high-energy electrons which, in turn, is a function of the acceleration voltage. The barrier can be treated with between two and seven Mrads before being plied up to the liner. Radiation treatment is typically applied only on the barrier side adjacent to the ply of the tire.

8.5 TIRE BUILDING

The third operation in tire manufacturing is tire building. As in component preparation, the tire building area has full temperature and humidity control, and, as in most industries, the trend toward increasing automation is being pursued.

Building is the process of assembling all the components onto a tire-building drum. As would be expected, the greater the degree of off-line preassembly of components

TABLE 8.3
Polymer Response to Electron Beam Radiation Processing (5, 6, 7)

Polymer and Compound Crosslinking	Polymer Degradation
Natural rubber	Bromobutyl
Styrene butadiene rubber (SBR)	Chlorobutyl
Polybutadiene	
Ethylene propylene diene rubbers (EPDM)	

TABLE 8.4
Building Machines

Type	Nominal Output (PCR)[1]	Advantages	Comment
Two-stage	300 units/shift	Most flexible for building complex construction	Two operators
Single-stage	400 units/shift	Efficient	Single operator
Automated	500 units/shift	Limited flexibility. Requires efficient component supply line	One operator for multiple machines.

[1] Theoretical output rates are much higher. 100% up-time is a function of adequate component supply, curing capacity, and maintenance support.

achieved, the greater will be the output of the building machine. Off-line assembly of components occurs when, for example, two components, such as a sidewall and a shoulder wedge, are co-extruded, using a duplex extruder, or three components, such as a sidewall, shoulder wedge, and apex #3 (outside ply ending in truck tires), are co-extruded, using a triplex extruder. Belt edge barriers applied at the belt building area, rather than at the tire building machine, would be another example.

There are essentially three types of tire building machines: two-stage machines, where the casing and belt and tread assembly are at two different stations, single-stage machines, and fully automated machines (Table 8.4).

The tire building operation can be described as follows:

1. First stage operation:
   a. The inner liner and barrier (squeegee or tire gum) is applied to the circular building drum off a server
   b. The casing ply or multiple plies are next applied onto the drum
   c. Beads, sidewalls, and chafer are wrapped around the drum or placed on the drum
   d. The turn-up or first shaping operation then follows, where the assembly is turned up over the bead.
2. Second stage operation:
   a. The carcass of the tire is inflated and the belts and tread are applied to produce the "green" (uncured) tire, the term "green" being used in the industry to describe an uncured or un-vulcanized compound or tire (Figure 8.12).
   b. Stitchers then come in to ensure all of the components are placed tightly together and trapped air is removed, after which the tire is inspected and weighed
   c. The tire is removed from the building machine and sent to curing

All compounds require splicing, including i) overlap splices, ii) butt splices, iii) tapered splices, or iv) where a splice is avoided, a "double wrap" can be used (Figure 4.13). A double wrap can be used for inner liners, where two layers of inner liner

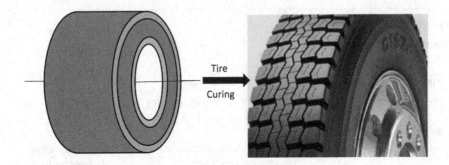

FIGURE 8.12· Schematic of a "Green" Tire

are applied at the building drum, thereby eliminating a splice and ensuring a more uniform gauge around the inside of the cured tire.

Tapered splice angles should be as low as possible, thereby maximizing the contact area in the splice and, in turn, maximizing resistance to any forces pulling apart the splice. Heavy overlaps should also be avoided as this can cause uniformity concerns, the heavy splice generating a harmonic vibration in the vehicle when travelling at high speeds. Splices that are too heavy or non-symmetrical will generate defects in force variation, balance, or bulge parameters. The 1st harmonic, typically assigned to the tread splice, can be constrained by off-setting, for example, the sidewall splices, thereby cancelling out any force vectors caused by a heavy splice.

Belts are spliced end-to-end with no overlap. Splices that are too light or open can lead to visual defects and, in some cases, tire defects, leading to warranty concerns.

There are four global tire building machine manufacturers, VMI in the Netherlands, HF from Germany, Mitsubishi in Japan, and Mesnac from China. In addition, the largest tire makers have developed automated tire-assembly machines in-house, which create potential advantages in terms of tire construction uniformity and production output rates. In some instances, labor requirements are reduced, but the "back-room" engineering requirements can be greater, due to increased operations complexity.

8.6 TIRE CURING (VULCANIZATION)

The fourth operation or stage in tire production is curing. This is the most capital-intensive operation in tire production (Figure 8.13). It therefore requires research and development work to ensure the operation is performed efficiently and that energy conservation is optimized.

The curing operation in a tire factory consists of two areas, the green or uncured tire storage or holding area, and the press area. Green tire storage is a critical area in which uncured tires could be held for up to several shifts before moving to the press line. Two conflicting variables are important here:

1. The green tire storage area should be at a controlled temperature. The higher the temperature of the green tire, the shorter will be the tire cure time. Being the most capital-intensive stage in the plant means that the

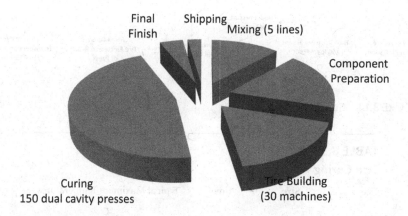

FIGURE 8.13 Investment Cost Distribution for a PCR Tire Plant (Based on a nominal construction cost of $525 million)

cumulative impact of several seconds of cure time reduction can have a significant impact on total plant output rates

2. The room temperatures must be kept sufficiently low to ensure the components do not separate. For example, there is considerable residual force at the ply ending such that, if the green tire is too warm, there is a risk of components shifting and distorting, eventually causing ply-end separations.

Optimum environmental controls in the green tire storage area will vary among tire companies because of differences in tire construction and compound. Process capability studies would therefore be needed to identify the target room conditions and tolerances so that, when the tire reaches the press, it is cured efficiently and meets the company's product production performance and quality control standards.

Curing is the process of applying pressure and heat to the green tire in a mold in order to give it its final shape. Heated molds facilitate the flow of the rubber in the green tire to achieve its final dimensions, tread pattern, and sidewall lettering, and then achieve vulcanization of all of the rubber compounds.

To elaborate, in this process, the green tire is automatically transferred onto the lower mold, a rubber bladder is inserted into the green tire, the upper dome and top part of the mold with the tread segments closes, while the bladder inflates (8). When the mold has closed and locked, the bladder pressure increases so that the green tire flows into the mold, taking on the tread pattern and sidewall lettering engraved in the mold sidewall plates. The bladder is filled with a recirculating heat transfer medium. Two systems are used, namely i) high-pressure steam at up to 200°C, followed by high-pressure hot water, or ii) high-pressure steam at up to 200°C, followed by nitrogen.

Cure cycles can be quite complex, with both steam and water or nitrogen entering the bladder at different stages and temperatures. In addition to the bladder, both the mold sidewall plates and tread segments will be heated. In the case of the bladder, the temperature cycle would follow a general sequence which is illustrated in Figure 8.14.

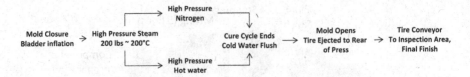

FIGURE 8.14   Cure Cycle Steps

TABLE 8.5
Tire Curing

Tire	Size	Time	Typical Maximum Temperatures
			°C
Passenger	205/55R16	9–16 min	190
TBR	275/80R22.5	35–46 min	150
OTR	40.00R57	48 h	120

Temperatures and cure times will vary according to the tire line. Table 8.5 shows some typical times for various types of tires, which vary considerably among tire manufacturers. In any case, the critical variables for optimizing cure time will be:

- Press calibration
- Steam quality (i.e., pressure consistency and no in-line condensation)
- Tire component gauge consistency and uniformity
- Bladder gauge and thermal conductivity
- Curing medium, with nitrogen being preferred when available
- Tire compound vulcanization systems

The cure time for passenger tires varies among the tire companies. For example, the cure time of a tire, size 205/60R16, can range from 9 minutes to 16 minutes, with maximum cure temperatures varying from 175C° to 190°C. Given the capital investment in tire curing capacity, those companies that have made the appropriate investments enjoy significant savings in both energy and equipment utilization.

Like many other parts of the tire industry, the global press manufacturers have been consolidated into four major producers, Larsen & Toubro, Mitsubishi, HF Group, and McNeil & NRM. Regardless, all modern tire presses consist of several main parts

1. Tire mold and dome or chamber enclosing the mold within which the tire is cured
2. Hydraulic systems to open and close the mold. The upper dome and mold of the press can be raised in either a tilt-back manner or vertically
3. Steam nitrogen and high-pressure hot water systems, depending on the curing medium used

4. Control systems for temperature sequencing, pressures, and time
5. Automated green tire loading and cured hot tire handling systems
6. Curing bladder, which inflates when the tire mold closes, forcing the green tire into the mold and forming the final tire shape, tread patterns, sidewall, and inner liner appearance

ExxonMobil has published two engineering manuals on curing bladders to which further reference is recommended (8,9) Briefly, the curing bladder is a cylindrical bag of specially compounded butyl rubber containing a poly-methylolphenol resin cure system. There are two types of bladders, one with both ends open, and one with one end closed. The collapsible bladder is mounted in the lower section of the tire curing press and forms a part of the press and mold assembly. The green or unvulcanized tire is positioned over the bladder in the bottom half of the mold. When the mold is closed, pressurized steam, air, hot water, or inert gas (nitrogen) is introduced systematically (in a pre-programmed cycle program) into the bladder to provide internal heat and pressure for the tire shaping and curing process. The formulation of a curing bladder compound is illustrated in Table 8.6.

At the end of the cure cycle, the pressure is released, there is a cold-water flush, the mold opens, and the tire is lifted out of the mold. Tires with a nylon ply may be placed on a post-cure inflator (PCI) which re-pressurizes the tire until it cools. The tire is then transferred to Final Finish.

Cure cycles tend to be complex, with pressure and temperature settings of the dome, mold, and bladder varying to achieve the optimum cure state for the tire. The cure cycle strategy varies among manufacturers due to differences in compound line-up, tire construction and component gauges, press design, and type and quality of the cure medium. A simple example of a steam–high pressure hot water cure cycle for a radial truck or bus tire is described below:

1. Steam is injected into the dome and the temperature reaches 155°C (tread segments at 155°C, sidewall plates at 150°C)
2. Bladder cycle: Steam inflates the bladder to reach 200°C. Steam pressure would be up to 16 kg/cm^2 (200 psi) followed by 22 kg/cm^2 hot water and then a cold-water flush. A cure cycle for a TBR tire, for example 275/80R22.5, would be:
   1. Steam   12′00″ - 13′00″
   2. HP hot water   30′00″ – 32′00″
   3. Cold water flush   3′00″ – 5′00″
   4. Drain   0′30″

Nitrogen cure cycles tend to result in better tire curing bladder service life and better-controlled temperature profiles. Equipment maintenance is easier and shows greater press up-time (Table 8.7).

The typical or average N_2 consumption in normal cubic meters (Nm^3 measured at 1 atmosphere and 0°C) for a passenger car tire cure period is around 0.50 to 0.55 Nm^3/tire and between 2.6 and 4.0 Nm^3/tire for a heavy truck or bus tire. These approximations suggest 0.7 kilograms of N_2 is used per cure. For a plant producing

TABLE 8.6
Effect of Butyl Rubber Isoprene Content and Curing Resin Type (9)

Compound		1	2	3	4	5	6
EXXON Butyl Grade 065		100.00			100.00		
EXXON Butyl Grade 165			100.00			100.00	
EXXON Butyl Grade 365				100.00			100.00
Polychloroprene (W Type)		5.00	5.00	5.00			
Carbon Black, N330		50.00	50.00	50.00	50.00	50.00	50.00
Castor Oil		5.00	5.00	5.00	5.00	5.00	5.00
Stearic Acid		1.00	1.00	1.00	1.00	1.00	1.00
Zinc Oxide		5.00	5.00	5.00	5.00	5.00	5.00
SP1045		10.00	10.00	10.00			
SP1055					10.00	10.00	10.00
Mooney Viscosity (ML1+8)		32.0	32.0	33.0	32.0	32.0	33.0
Isoprene (mol%)		1.05	1.50	2.30	1.05	1.50	2.30
Curing Resin		10.45	10.45	10.45	10.55	10.55	10.55
Mooney Viscosity ML1+4 (100°C)	[MU]	59.8	60.0	63.3	63.2	63.9	66.7
Rheometer (MDR) (180°C)							
MH-ML	[dNm]	5.4	6.8	9.3	3.6	4.7	6.7
t10	[Min]	1.5	1.9	2.7	1.0	1.0	1.0
t90	[Min]	25.9	25.7	25.1	19.1	14.7	9.8
Tensile Strength	[MPa]	11.66	13.44	13.86	12.08	12.07	13.20
Elongation	[%]	690	650	520	750	770	700
300% Modulus	[MPa]	4.17	5.74	8.05	3.43	3.60	5.72
7 Days at 125°C	[MPa]	11.42	11.67	12.35	11.07	11.09	11.40
Elongation	[%]	550	500	370	680	650	500
300% Modulus	[MPa]	7.16	8.18	10.84	4.64	5.28	7.40
Steam-Aged 96 hours at 180°C	[MPa]	11.48	11.11	12.16	11.10	10.97	10.42
Elongation %	[%]	490	400	260	660	600	400
300% Modulus	[MPa]	6.94	8.83	-	3.93	4.91	7.95
Hardness Shore A	[Shore A]	60	61	61	62	60	59
7 Days @ 125°C	[Shore A]	81	79	80	76	75	76
Steam-Aged 96 h @ 180°C	[Shore A]	86	85	86	69	71	74
Tear Strength Die B	[N/mm]	57.21	65.66	54.74	54.03	62.38	58.74
3 Days @ 125°C	[N/mm]	59.95	60.92	40.67	63.03	59.18	51.34
7 Days @ 125°C	[N/mm]	54.85	49.04	40.96	47.27	49.60	42.92
Steam Aged 96hrs @ 180°C	[N/mm]	45.22	45.56	34.70	47.70	47.89	33.49
Tension set	[%]	15.4	9.7	6.3	25.3	17.5	9.6
Fatigue to Failure	[K-Cycles]	269.4	161.7	47.4	269.6	266.6	141
Permeation Coefficient	cc ×mm/(m²-day)	169	173	175	143	144	154
Permeability Coefficient	cc ×mm/ (m²-day-mmHg)	0.251	0.257	0.260	0.212	0.215	0.23

TABLE 8.7
Simplified Nitrogen Cure Cycle for a Commercial Truck Tire (TBR)

Step	Time Start	Duration (minutes)	Description	Maximum Attained Temperature
1	0 min	0.5	Close mold, bladder inflates	
2	0.5 – 12.0	11.5	200 psi steam in bladder (bladder temperature 180–185°C)	185°C
3	12.0 – 39.5	27.5	Nitrogen	
4	39.5 – 40.0	0.5	Cold water flush	
5	at 40 minutes	0.0	Tire ejected	

25,000 PCR units/day, this is approximately 16 to 17 tons of liquid nitrogen per day, assuming no recycling.

Nitrogen cure systems i) allow directionally lower cure temperatures with energy savings, ii) have better control of temperature during the cycle, and iii) tend to result in better bladder life due to the absence of any degradation due to steam or hot water. Keeping nitrogen dry is as important as preventing water condensation in steam lines. Condensation in molds typically causes an irregular cure state between the bottom serial side and top non-serial side of the molded tire

A final subject area in tire curing is molds. Tire molds come in two basic types, 2-piece and segmented (typically, seven segments or more). Segmented molds are preferred, though 2-piece molds are cheaper. Molds would be acquired in sets or "complements", depending on the number of tires to be built and the number of tire machines committed to building a specific tire. For example, if the output rate of a building machine is 400 tires per shift, then there must be sufficient curing capacity to take all the tires. If the cure time is 12 minutes, including loading and unloading, one mold can cure 40 tires in one shift. To cure all the tires coming off one machine, ten molds will be required, plus one spare. Therefore, a complement in this case would be eleven molds. To cure a tire design and size, molds would be acquired in sets or complements of eleven, and the number of complements would depend on the number of building machines allocated to the tire construction.

8.7 FINAL FINISH

After the tire has been cured, there are several additional operations. After curing, the hot tire is ejected from the mold and transferred by a conveying system to an inspection station in the Final Finish area of the plant, which represents the fifth manufacturing operation. This area covers:

1. Visual check by an inspector for defects, such as incomplete mold fill, exposed cords, blisters, and blemishes

2. Removal of any flash, if present
3. Uniformity measurements
4. X-ray, Shearography, balance, and any other special inspections required by a customer for specific tires
5. Removal and transfer of tires to the laboratory for i) cut-tire analysis and component gauge measurements, ii) durability testing to FMVSS 139 (PCR) or FMVSS119 (TBR), and iii) other specific statistical quality control (SQC) tests that the plant has in place.

The actual number of inspection stations will be determined by the manufacturer's quality assurance policies and the outcome of statistical quality control studies. In addition, uniformity measurements will be obtained, i.e., tests where the tire is mounted on a wheel, inflated, run against a simulated road surface, and measured for force variation. Cut-tire gauge checks should be conducted for each tire size and type in production once per week or more frequently, depending on the tire line and the requirements of the end-user, such as the car manufacturers.

Large commercial truck/bus tires, as well as some passenger and light truck tires, are inspected by X-ray which can penetrate the rubber to analyze the steel cord structure, cord placement uniformity, and check for any unraveled wire filaments.

8.8   SHIPPING

Depending on the location and company policies, plants can have different methods for handling shipment of final tires. Examples include:

1. A warehouse attached to the plant and then direct shipping to a dealer or distributor
2. Shipping as a staging area for bulk shipments to distributors, who then handle the sorting and subsequent distribution to dealers or original equipment manufacturers
3. Sorting and then shipping to a company-owned warehouse, followed by distribution to company-owned stores, independent dealers, or original equipment manufacturers,

In any case, adequate floor space is needed to handle the flow and distribution of products from the manufacturing area of the plant.

8.9   TIRE PRODUCTION AND THE ENVIRONMENT

Like all industrial operations, tire manufacturing should be structured to minimize its environmental footprint. There are three important aspects to this, namely renewable sources of raw materials with adequate shelf life, post-industrial waste recovery and disposal, and emissions.

Renewable raw materials include natural rubber, discussed earlier, naturally occurring resins such as pine tars, and novel fillers, waxes, and fatty acids. Many tire companies have demonstrated new tire constructions using biomaterials, and the final product performance was very competitive when compared with tires built from more conventional materials.

At an operational level in a manufacturing facility, the focus on raw materials and environmental controls would include,

1. Minimum emissions, and, where volatile organic compounds (VOC) do come off hot compounds, appropriate facilities are in place to capture the VOCs and treat them (e.g., scrubbers)
2. No dust, i.e., dust-free powders. Raw materials are received at the manufacturing plant as pellets (clay carrier) or pastilles (wax carrier). Some materials, such as silane coupling agents or components in reinforcing resins, can come in liquid form, allowing them to be stored in tanks and then automatically injected into the Banbury or internal mixer when compounds are being mixed
3. No odor
4. Elimination of process oils containing polycyclic aromatic hydrocarbons (PAH), i.e., high aromatic content oils
5. Elimination of toxic materials, such as chromium, mercury, lead, minimizing of zinc and alpha quartz, found in some talc powder
6. Use of non-nitrosamine-generating accelerators
7. Auto-weighment systems for chemicals, sometimes referred to as preweigh, and packaging in heat-sealed bags, thus eliminating free powders in the compound mixing area, with benefits for industrial hygiene
8. No free carbon black or silica, i.e., have the fillers in silos or in sealed one-ton "super-sack" systems
9. Post-industrial waste or compound recycling. As part of the compound development process, "scrap compound work-away" policies would be in place to help ensure attainment of as close to 100% conversion of compounds into final products as possible.

In addition to raw material controls, ergonomic safety systems, environmental noise minimization, appropriate lighting, air conditioning, and automatic movement of materials and tire components will help ensure a safe and productive work environment.

One of the most important elements in tire plant environment protection is minimization of compound and component waste. In many instances, some scrap generation is inevitable but, when it occurs, a root–cause analysis is necessary. An important area to focus on is the specification for a minimum compound shelf life which covers the period from mixing to extrusion or calendaring. Each compound released to manufacturing will have this as part of the release specification, along with mixing protocols and compound mechanical property specifications. Many techniques can be used to extend the shelf life of a compound tire factory, such as monitoring of lay-down temperatures at the end of the compound mixing line, slurry or anti-tack control, use of constant viscosity polymers, peptizers, retarders, and production planning and materials control policies. Equipment calibration is equally important to prevent generation of off-specification components.

There has been a recent trend to automate many of the operations in tire manufacturing. Equally important is improving energy efficiency in manufacturing, where curing is the most energy-intensive step. There are four areas to potentially improve curing times and increase curing operation throughput rates,

1. Component gauge optimization, consistency, and tire weight uniformity
2. Nitrogen gas in place of high-pressure hot water systems
3. Curing press calibration and press cure cycle tuning, enabling temperature reductions with no increase in cure time
4. Compound vulcanization system development.

Of these four items, the first three can be addressed over a reasonable period. Item four would require more extensive tire testing but the net gain will generate appropriate returns.

Regarding the final product, there are a number of items that can lead to longer product life, increased operating safety, and greater end-user satisfaction. The easiest to implement is a fully, 100% bromobutyl compound inner liner. It is well documented that a 1-mm gauge inner liner compound, with 100 phr of bromobutyl extending bead to bead, will lead to increased tire durability, tread wear, traction due to a better tire footprint, and lower vehicle fuel consumption. From the perspective of the manufacturer, the improved durability could lead to lower warranty claims. This is similar for truck and bus tires. A tire with an inner liner compound containing 100 phr of bromobutyl, extending from bead to bead and with a 2.0-mm gauge, will demonstrate better durability, lower adjustments, and allow improved vehicle fuel consumption due to tire inflation pressure retention.

8.10   SUMMARY

Tire manufacturing is a complex operation involving the processing of many types of raw materials, production of many different calendared and extruded profiles, and assembly of a final uniform composite. It is easy to benchmark tire manufacturers' capabilities, and measure the dimensions, properties, and performance of the final product. Compound formulations used by different manufacturers tend to be more difficult to benchmark but have a substantial impact on the specifications to be followed by the manufacturing organization. Curing is the most capital-intensive part of the tire plant and, on viewing the global tire manufacturing base, potentially offers the largest area for future improvements in manufacturing efficiencies and energy savings.

REFERENCES

1. Bhowmick AK, Hall MM, Benarey HA. *Rubber Products Manufacturing Technology*. CRC Press, Marcel Dekker, New York. 1994.
2. ELL Technologies. *Tire Manufacturing Data*. Cedar Park, TX. 2020.

3. Ninjmam G. Recent Developments in Rubber Mixing and Cord Calendering in Tire Production. Chapter 35. In *Current Topics in Elastomer Research*, Ed. AK Bhowmick. CRC Press, Boca Raton, FL. 2008.

4. Pohl JW, Limper A. Mixing Machinery for the Rubber Industry. In *Rubber Products Manufacturing Technology*. Eds. AK Bhowmick, MM Hall, HA Benarey. CRC Press, Marcel Dekker, New York. 1994.

5. *Exxon Halobutyl Rubber Tire Innerliner Processing Guidelines*. ExxonMobil Chemical Company, Houston, TX. 2011.

6. Drobny JG. *Electron Beam Processing of Elastomers*. Presented at a Meeting of the American Chemical Society Rubber Division, San Antonio, TX. 2005.

7. Thorburn B. *Effective Process for Pre-curing of Tire Components*. Presented at a Meeting of the American Chemical Society Rubber Division, San Francisco, CA. 2003.

8. *Tire Curing Bladder Technology Manual*. www.exxonmobil.com. ExxonMobil Chemical Company, Houston, TX. 2008.

9. Rodgers B, Jacob S, Curry C, Sharma BB. *Butyl Rubber Curing Bladder Resin Vulcanization Systems: Compositions and Optimization*. Presented at a Meeting of the American Chemical Society Rubber Division, Pittsburgh, PA. 2009.

9 Manufacturing Quality Control and Tire Uniformity

9.1 INTRODUCTION

Tire quality is the most important differentiator among the global tire companies. Typically, there are five inputs to meeting quality targets, and these are:

1. Purchasing: Dependability of supply, regulatory compliances, and pricing
2. Technology: Development of the raw material specifications, the manufacturing specification, and tire component specifications, and specification compliance
3. Manufacturing: Meeting the production specification in an efficient manner
4. Quality Assurance: Monitoring, regulatory compliance, consistency, and uniformity
5. Final Product Performance: Laboratory tire durability, rolling resistance, and any specific original equipment vehicle manufacturers' requirements

The structure of the raw materials purchasing specification has been outlined earlier. This purchasing specification document should be prepared by the company's Technology Department, along with compound formulations and processing parameters. The Technology Department will also prepare the manufacturing specification, which will include tire component dimensions, gauges, and weights, the tire component building sequence, and final tire dimensions, including weights. Quality Assurance which, in some companies, is integrated into the Technology Department, will monitor the manufacturing process for compliance with the production speciation. Bhowmick et al. prepared a text covering many of the manufacturing processes involved in the rubber industry, including the importance of quality assurance (1). However, there is a gap in the literature pertaining to quality control in tire production, such as identifying critical quality control points in each of the base operations, such as mixing or extrusion. The discussion in this chapter is therefore intended to provide a starting point for consideration when planning a quality program in tire manufacturing.

Tire manufacturing is divided into six unit operations, within each of which there will be quality control points, namely raw materials receiving and mixing, component preparation, building, curing, inspection, and shipping (Figure 9.1).

Quality Control Points in Plant Operations

1	2	3	4	5	6
Raw Materials Receiving & Mixing	Component Preparation	Tire Building	Curing	Inspection Quality Control Laboratory	Warehouse Shipping
Quality Control Points: 1. Raw materials inspection 2. Raw materials compliance to purchase specification 3. Compound testing (RPA, viscosity, curing, mechanical properties)	Quality Control Points: 1. Extrusion checks (gauges, tack, weight contours, through put rates) 2. Calendering checks (gauges, tack, weight, through put rates	Quality Control Points: 1. Machine calibration 2. Tire weights	Quality Control Points: 1. Press calibration 2. Steam quality 3. Temperature controls	Quality Control Points: 1. Visual inspections 2. Uniformity 3. Laboratory durability and other tests to FMVSS139, FMVSS119	Quality Control Points: 1. Shipping order compliance

FIGURE 9.1   Manufacturing Unit Operations Quality Control Points

9.2   RAW MATERIALS AND COMPOUND MIXING

In raw materials receiving, quality assurance will include the visual inspection and the shipment-receiving tests outlined in the purchase specification. For example, in the case of polymers, it might include Mooney viscosity, moisture or volatile content, and ash content. In some instances, compounding in a model formulation to obtain vulcanization data may be requested. This might occur for halobutyl polymers, which are highly saturated, carbon black grades, where surface chemistry can influence final compound vulcanization kinetics, or accelerators, when the supply source has changed.

In most tire and industrial product manufacturers, mixed compound certification testing consists of two processes, namely baseline testing used for monitoring long-term trends, and batch certification.

The frequency of batch testing and certification will be determined by the statistical process controls and capability studies established by the company. For example, based on such statistical studies, and considering the case of a 100-batch run of a compound in a large tire plant, not every batch might need to be tested. Frequently, testing of the first series of batches, those from the middle of the run, and batches from the end of the run might be sufficient, provided there are no process deviations, such as batch weights, mix times, mix cycle temperatures profiles, or other internal mixer operating settings.

Such process controls require continued monitoring, with the objective of keeping the compound properties at the center or mean of the specification. Over time, should the average data drift off the mean, an alert limit or other early "attention" point should be activated, at which point a root–cause analysis should be initiated to identify the cause of the drift. The alert limit could be set at mean $\pm 1\sigma$ (standard deviation), so that the material is still well within specification set by earlier quality standards (Figure 9.2). Upper and lower control limits can be set, using data to calculate mean $\pm 3\sigma$, plus a coefficient determined by the allowed tolerances, ensuring that tires meet their design performance (Figure 9.3). This would be the Upper and Lower Specification Statistical Limits (USL and LSL). Regardless of how upper

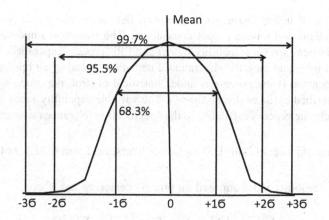

FIGURE 9.2 Setting Upper and Lower Control Limits by Measurement of Standard Deviation for a Data Set

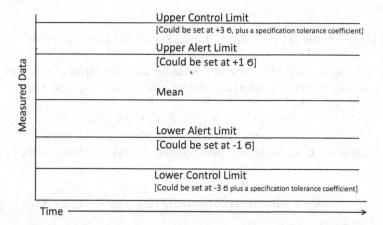

FIGURE 9.3 Upper and Lower Control Limits; Standard Deviations Used to Define Alert and Control Limits

and lower control limits are set, and no two companies are the same, the alert limit should, in all cases, be the trigger point for corrective action to address data drift.

In compound quality assurance, the trend in recent years has been to use the Rubber Processability Analyzer (RPA) test equipment. This instrument, initially based on the moving die rheometer (MDR), provides a lot of very useful information on the compound, including viscosity, vulcanization kinetics, compound rheology, such as die swell, and cured properties, including storage and loss modulus. In batch quality testing, it is conducted at between 190°C and 200°C, thereby enabling the test to be completed in around 4 minutes.

Baseline testing is more extensive and may include testing for Mooney viscosity, the RPA protocol of testing, and compounding to obtain vulcanization data, tensile strength, and resilience, such as rebound and dynamic mechanical testing analysis (DMTA).

A number of indices have been developed that allow plant quality control to monitor systems and ensure output consistency. The two most important process capability indices are Cp (capability index) and PPI (process capability), which are determined by calculating the six standard deviation spread of an operation under statistical control. If the process is under statistical control, the process capability can be quantified by determining the 6σ. The variable capability index (Cp) is the means of relating process capability to the specification tolerances where:

$$Cp = \left[\text{Upper Statistical Limit (USL)} - \text{Lower Statistical Limit (LSL)}\right] / 6\sigma \qquad (9.1)$$

CPC is the capability index adjusted for process centering and where

$$CPC = \left[\text{USL} - X\right] / 3\sigma \text{ or } \left[\text{LSL} - X\right] / 3\sigma \qquad (9.2)$$

The Process Performance Index (Pp) relates the performance to the specification tolerances

$$Pp = \left[\text{Upper Statistical Limit (USL)} - \text{Lower Statistical Limit (LSL)}\right] / 6s \qquad (9.3)$$

where "s" is an estimate of the standard deviation calculated from sample data, PPP is the process performance index, adjusted for process centering, and where

$$PPP = \left[\text{USL} - X\right] / 3s \text{ or } \left[\text{LSL} - X\right] / 3s \qquad (9.4)$$

The equation for Cpk, which is defined as the process capability index is more complex:

$$Cpk = \text{minimum} \left\{\left(\text{USL} - \text{mean} / 3\sigma\right), \left(\text{mean} - \text{LSL} / 3\sigma\right)\right\}$$

Often, Cpk is also described as the capability a process is achieving, whether or not the mean is centered between the specification limits.

Cpk represents the lowest value of the capability against the upper or lower specifications, showing where, within the specification limits, the process is producing. To calculate the Cpk measurement, one compares the average of the data to the specification limits. A Cpk that is less than 1.33 needs some action to make it higher, whereas a Cpk of less than 1.0 means that the process is not capable of meeting its requirements. A large number of texts have been published on quantifying manufacturing quality data, and further reference to such work is strongly encouraged.

9.3 COMPONENT PREPARATION

Quality assurance in this area is focused on component dimensions, tolerances, and weight. Most emphasis would be placed on equipment calibration, such as temperature controls on extruders, creel room environmental conditions, and water-cooling systems on extrusion lines and tension units on calendaring and extrusion lines. This

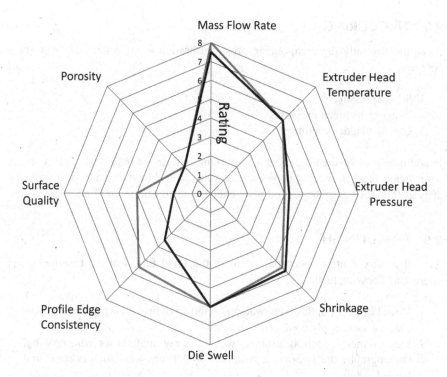

FIGURE 9.4 Variation in Processability between Two Polybutadiene Polymers Purchased under the Same Specification (2)

area has an important link to the raw materials purchasing specification, in that variations due to source of supply can be reflected in the processability of the compounds (Figure 9.4). Tire factory process variability, attributable to two different raw material suppliers, is handled in two ways:

1. Factories will single-source a raw material, thereby minimizing a potential source of process variability. When multiple suppliers are shipping the same product to a customer, e.g., a carbon black such as N660, they would be directed to be the sole supplier to a specific facility, and the competing supplier would be designated to supply to a different plant operated by the customer
2. Compound formulations should be sufficiently well developed so as to be desensitized to raw material anomalies, that, in turn, might induce process variability. This, in turn, would be achieved by the use of masterbatches, and process aids, such as peptizers, homogenization agents, and retarders.

9.4 TIRE BUILDING

The item of most importance in tire building is machine calibration. This includes server and drum alignments, laser light alignment, stitching pressures, and visual inspection of components as they are applied at the building machine. The final quality check at this point would be green tire weight.

9.5   TIRE CURING

As in the tire building area, curing press calibration is important. The variables here are:

1. Steam pressure
2. Water or nitrogen pressures
3. Curing bladder condition

In addition, press dome alignment is important. In terms of industrial hygiene, leaks from steam pipe, nitrogen gas pipes, and water pipe should be addressed immediately as this has a large impact on plant energy and consumption of make-up water.

9.6   FINAL FINISH

Several quality control operations occur in the Final Finish of the tire inspection operation. These include:

1. Visual inspection, flash removal, and removal of tires showing blemishes or defects, such as ply cord strike-through
2. Measurements, such as balance, weight, X-ray analysis for wire spacing, shearography (for internal separations), and dimensions, such as static and loaded radius
3. Laboratory testing, including i) cut-tire analysis, where tire sections are cut and component gauges measured to ensure compliance with the building specification, and ii) tire durability, high-speed, and strength tests
4. Uniformity

Of the four sets of quality control checks, uniformity is, in many instances, a key differentiator between tire manufacturers. First, considering each of the four focus areas, in visual checks, inspectors would be checking for the following blemishes:

a. Compound degradation due to scorch
b. Inner liner thin spots
c. Cracks
d. Blisters, particularly in the inner liner
e. Obvious separations, seen as localized bulges in the sidewall, bead area, or the inner liner
f. Off-center inner liner or other external components, such as the sidewall
g. Splice integrity

After a visual inspection, data collected in Final Finish, such as balance, cut-tire gauge analysis, and weight, and from non-destructive testing, such as shearography, then follows. Shearography has proved to be most effective in identifying internal separations. Essentially, a tire is placed in a chamber, a vacuum is applied, and irregular surface patterns corresponding to a separation can then be identified (Figure 9.5).

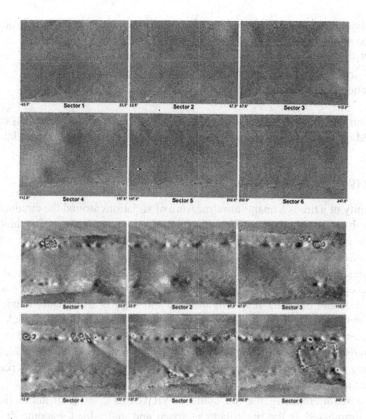

FIGURE 9.5 Shearography patterns: None *versus* Many Separations (3)

New tires can be checked, either on a 100% basis or following a statistical protocol, which, in turn, would be based on a capability study. In truck tire retreading today, nearly 100% of tire casings are checked before proceeding to the retread operation.

Quality control laboratory testing covers two principal areas, gauge analysis of dissected tires and durability testing, according to a recognized standard such as the United States Federal Motor Vehicle Safety Standards (FMVSS) 119 and 139.

This would also include such tests as rolling resistance for OEM tires. Laboratory testing of tires will be discussed in Chapter 10, on Tire Testing.

Gauge analysis of cut-tire sections could be done on a weekly basis or once every two weeks, depending on the tire line and statistical data collected on component placement variability. In the case of tires which are always in production, such as the 275/80R22.5 steer axle truck tire, a weekly check could be conducted to set tolerances and to evaluate variability, after which evaluation frequencies could be reset. The critical gauges to be monitored will change with each manufacturer, but some control points could be:

1. Belts and overlay placements (i.e., width and being on-center)
2. Tread and base centerline gauges, base and tread shoulder gauges, and inner liner to tread

3. Shoulder and wedge gauges (including from the inner liner to the outside of the shoulder)
4. Ply-end height
5. Sidewall gauges (from inner liner to outside sidewall)
6. Inner liner gauge uniformity

Up to fourteen data points could be collected and then, with time, a process capability could be established, from which improvement programs can be formulated

9.7   TIRE UNIFORMITY

Uniformity of a tire is a quantitative measure of variations around the circumference of a tire. It is also the greatest differentiator among tire manufacturers and thus merits attention in manufacturing. Uniformity variations are measured as forces when the tire rotates. Tire uniformity describes the dynamic mechanical properties of pneumatic tires that have been defined by a set of measurement standards and test conditions accepted by global tire industry and automobile manufacturers. As passenger vehicle ride and handling standards increase and there is also greater emphasis on truck handling and comfort, uniformity standards will become tighter. Vehicle manufacturers, however, do not have a common set of standards, and due to the dampening characteristics of different vehicles, a tire which performs well on one vehicle platform may not necessarily perform equally well on a different platform. It is pertinent, however, to define the range of terms used to define tire uniformity.

The circumference of the tire could be viewed as a series of anomalies that, when compressed as the tire rotates at speed and under load, become deformed as they enter the road contact area, and recover as they exit the tire's footprint. Typically these anomalies are created by either heavy or light splices, component gauge variation, and compound misplacement. Variation in both radial and lateral directions cause variations in the compressive and restorative forces as the tire rotates (Figure 9.6). Given a perfectly symmetrical tire operating on a uniform roadway, forces created between the car and the tire will be insignificant. However, typical tires operating on a typical highway will exert varying forces into the vehicle that will repeat at every rotation of the tire. This variation is the source of various ride disturbances and driver discomfort, and both tire and car manufacturers work to reduce such vibrations, this being necessary to improve the dynamic performance and comfort of the vehicle.

9.8   FUNDAMENTAL TIRE FORCES

Tire forces, illustrated in Figure 9.6, have been reviewed by Ford and Charles (4) and Rodgers and Waddell (5) in much detail, and to which further reference is recommended. Tire forces are divided into three axes, radial (Fz), lateral (Fy), and tangential (Fx). Fz is the vertical force between the tire and the road, running from the road to the tire center, and to the tread and the vehicle. This axis supports the vehicle's weight and load. The lateral axis, Fy, runs sideways on the rotation axis. This axis is

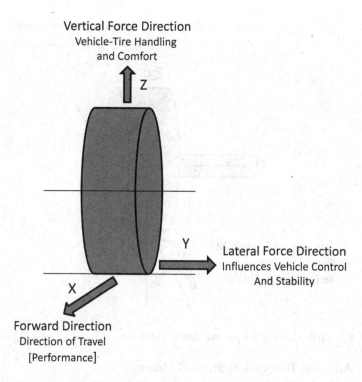

Vertical Force Direction
Vehicle-Tire Handling
and Comfort

Z

Y

Lateral Force Direction
Influences Vehicle Control
And Stability

X

Forward Direction
Direction of Travel
[Performance]

FIGURE 9.6 Fundamental Force Vectors Acting on a Rotating Tire (Axes of Measurement)

parallel to the tire mounting axle on the vehicle and applies to vehicle steering. The tangential axis, Fx, is the one running in the direction of tire travel. It is the driving force as applied to the tire.

9.8.1 RADIAL FORCE VARIATION

Radial force variation describes the change in this force as the tire rotates under load. As the tire rotates and spring elements with different spring constants enter and exit the contact area, the force will change. Consider a tire supporting a 500 kg load running on a perfectly smooth roadway. It would be typical for the force to vary up and down from this value. A variation between 498 kg and 502 kg would be characterized as an 4 kg radial force variation (RFV). RFV can be expressed as a peak-to-peak value, which is the maximum minus the minimum value, or any harmonic value, as described below.

Many tire manufacturers mark the sidewall with a red dot to show the location of maximal radial force and runout, and this is referred to as the high spot ("high spot marking"). A yellow dot marks the point of least weight. Use of the dots is specified in the Technology Maintenance Council's performance standard RP243. To compensate for this variation, tires are supposed to be installed with the red dot near the valve stem, assuming the valve stem is at the low point, or with the yellow dot near the valve stem, assuming the valve stem is at the heavy point (4).

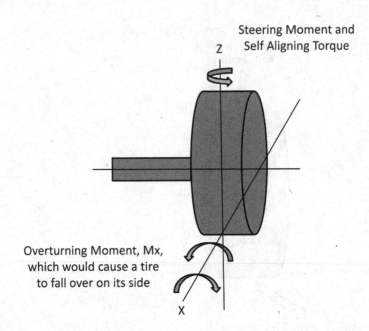

FIGURE 9.7   Self-Aligning Torque and Steering Moment

9.8.2   ALIGNING TORQUE AND STEERING MOMENT

This is the force that produces a steering pull and is the torque that aligns the steering back to the center after a turn (Figure 9.7).

9.8.3   OVERTURNING MOMENT (Mx)

This is the force that would tend to make a tire fall over on its side and can induce camber (Chapter 10). Camber is the tilt of the front wheels on a vehicle. Outward tilt from the vertical axis, Z, is positive camber, with inward tilt at the top being negative camber.

9.8.4   RADIAL FORCE VARIATION (Fz)

This is the change in vertical force as an inflated and loaded tire is rotated (Figure 9.6).

9.8.5   RADIAL RUN-OUT

This is a measure of the out-of-roundness of a tire and is a measure of how far the surface of a tire tread centerline varies from a true circle, i.e., variation in the free radius (Figure 9.8). It can be expressed as the peak-to-peak value as well as harmonic values and will induce a vibration in the vehicle similar to radial force variation. Some tire manufacturers mark the sidewall with a red dot to indicate the location of maximal radial force and runout.

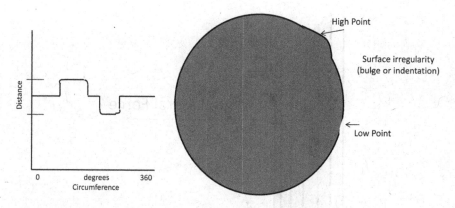

FIGURE 9.8   Radial Run-Out

9.8.6   LATERAL FORCE VARIATION

The lateral force, Fy, acts from side-to-side along the tire and wheel axle, and lateral force variation describes the change in this force as the tire rotates under load. High lateral forces can cause the tire to pull to one side or in one direction. For a large heavy-duty commercial truck tire, a variation between 10 kg and 12 kg would be characterized as a 2 kg lateral force variation, or LFV. LFV can be expressed as a peak-to-peak value, which is the maximum minus the minimum value, or any harmonic value, as described in Section 9.8.10. Lateral force is signed, such that, when mounted on the vehicle, the lateral force may be positive, making the vehicle pull to the left, or negative, pulling to the right.

9.8.7   CONICITY

This is a lateral force described as lateral shift clockwise plus the counterclockwise force, divided by 2. Qualitatively, the tire will roll like a cone with the vehicle pulling to one side (Figure 9.9). It is typically caused by off-center belts or misaligned belts. Conicity is an important parameter in production testing. In many high-performance cars, tires with equal conicity are mounted on left and right sides of the car in order that their conicity effects will cancel each other out and generate a smoother ride performance, with little effect on steering. This necessitates the tire maker measuring conicity and sorting tires into groups of similar values.

9.8.8   PLY STEER

This is a lateral force, defined as a lateral shift clockwise minus the counterclockwise lateral shift, divided by two (Figure 9.10). Ply steer describes the lateral force a tire generates due to asymmetries in its carcass as it rolls forward with zero slip angle and may be referred to as a pseudo sideslip. It is the characteristic that is usually described as the tire's tendency to "crab walk", or move sideways while maintaining a straight-line orientation. This tendency affects the steering performance of the vehicle. In order to determine ply steer, the lateral force generated is measured as the

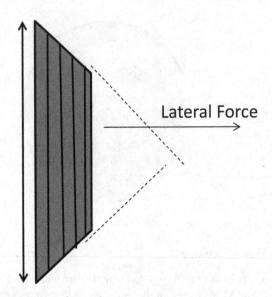

FIGURE 9.9 Conicity

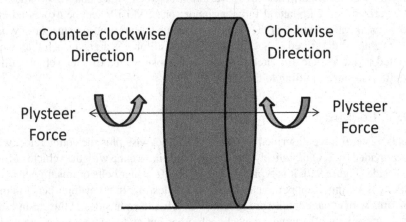

FIGURE 9.10 Ply Steer

tire rolls both forward and back, and ply steer is then calculated as one-half the sum of the values, keeping in mind that the values have opposite signs.

9.8.9 LATERAL RUN-OUT

This describes the deviation of the tire's sidewall from a perfect plane and is expressed as a peak-to-peak value around the sidewall circumference and is therefore analogous to radial runout. Lateral run-out can induce a vibration in the vehicle in a manner similar to lateral force variation (Figure 9.11). Lateral run-out is most often measured in the upper sidewall, near the tread shoulder.

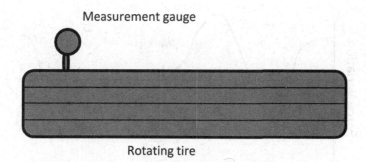

Variation in sidewall geometry while loaded and inflated tire rotates

FIGURE 9.11   Lateral Run-Out

Sidewall bulges and depressions can create this lateral run-out condition. A bulge is a weak spot, or a location of irregular gauge components, in the sidewall region that expands when the tire is inflated. Conversely, a depression is either a reinforced location that does not expand in equal measure into the surrounding area, or a location where irregular gauge components or splices are located. Bulges may also suggest construction defects, such as irregular cord spacing.

9.8.10   Harmonic Waveform Analysis

Force variation measurements can be represented as a waveform. The first harmonic, expressed as a radial force first harmonic (RF1H), describes the magnitude of the force variation that exerts a pulse into the vehicle once for each rotation. The first harmonic can be created by an irregular or heavy tread splice. The 2nd harmonic radial force, or RF2H, could be created by the sidewall splices, and RF2H expresses the magnitude of such radial forces that exert a pulse twice per revolution (Figure 9.12).

Often, these harmonics have known causes and can be used to diagnose production problems. For example, a tire mold installed with 8 bolts may thermally deform as to induce an eighth harmonic, so the presence of a high RF8H would point to a mold-bolting problem. RF1H is the primary source of ride disturbances, followed by RF2H. Historically, high harmonic values have not been a serious performance concern due to the frequency of the rotation of the tire at highway speeds. High vehicle speeds and harmonic frequencies could be damped or overcome by other vehicle dynamic conditions. However, as vehicle suspension systems become more complex and sensitive, such tire harmonic variations are becoming more important. Vector analysis will enable the engineer to position splices, and improve splice quality, around the circumference of the tire so as to cancel out excessive anomalies, thus ensuring a more uniform construction.

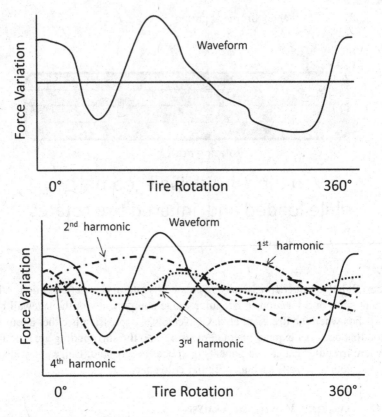

FIGURE 9.12   Harmonic Waveforms

9.8.11   GRINDING

A practice still found in some tire factors is grinding. This is where compound is removed to correct force variation anomalies. The result is higher production yields, less scrap, and less industrial waste. In the event that grinding is perceived to be necessary, a root–cause analysis should be initiated, bringing the manufacturing process back into specification, so as to eliminate the need for this operation.

In summary, a sample of data typically collected at a tire uniformity machine for a tire on a full-sized US automobile has been summarized in Table 9.1. Such data would typically not follow a Gaussian distribution but rather one with a longer tail, reflecting the number of tires deviating from the center of the specification (Figure 9.13). The intent is to minimize this tail but, in the event of an increase in off-specification production, a root–cause analysis is essential to bring the production quality back into line.

9.9   GLOBAL QUALITY MANAGEMENT STANDARDS

In the mid 1990's, a series of new quality standards were developed and most companies participating in the automotive industry have adopted them. The initial

TABLE 9.1

General Example of Tire Uniformity Values (Passenger car tire size P225/60R16 or tire of equivalent dimensions)

	Example of Data Range (in g)	
	Lower	Upper
Balance static	800.00	1000.00
Dynamic balance, non-serial side [upper control limit]	23.00	25.00
Dynamic balance, serial side [lower control limit]	28.00	32.00
Radial force variation	8.00	11.00
Lateral force variation	3.50	5.00
Conicity	0.50	1.00
Radial run-out	0.80	1.20
Lateral run-out	0.70	0.90
High point	0.35	0.45
Low point	0.25	0.35

governing standard is ISO9000 and its derivatives. The foundation of ISO 9000 is built on the company having defined quality systems, design, and development procedures, documentation, data traceability, calibration procedures, corrective and preventative action programs, audits and record-keeping procedures, and, finally, training. In total, over 22 elements make up the ISO 9000 series of standards. The standards have also undergone an evolution to QS9000 specifically for the large vehicle original equipment manufacturers and their Tier 1 suppliers, and then to TS16949, which focuses on additional factors, such as, for example, training, Tier 2 supplier audits, and labor force policies.

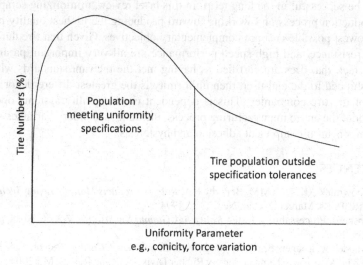

FIGURE 9.13 Nominal Tire Uniformity Distribution

At the Technology Department of a tire company, the product development process and many other internal projects have incorporated these three governing industry standards. In summary, the technology process now typically includes the following steps:

1. A project design plan, defining the end-product and its performance parameters
2. Design inputs, such as market data
3. Design outputs, including test data
4. Design reviews
5. Verification data obtained via prototyping
6. Validation *via* field trials
7. Design changes, if needed, followed by production release

Coupled with the discipline afforded by Stage-Gate systems, project road-blocks tend to be identified earlier and project success rates are typically higher. Most of these points fall into the many additional development tools, such as Kaizan, centered on the philosophy of continuous improvement, Total Quality Control (TQC) and Quality Function Deployment (QFD) management tools in which every technology leader should be competent. The tool, "Design for Six Sigma", has been the umbrella for many of these methods when used in new product development processes. There has been a variety of other technology management techniques that have been very successful. Though more qualitative in nature, management techniques such as "Management by Objectives (MBO)", "Employee [or Engineer] Engagement and Influence", and "Communications for Commitment" are all important elements necessary in any quality program.

9.10 SUMMARY

In conclusion, a high standard of quality of engineered product is essential for a business to be successful in the long term. In this brief review, minimizing scrap in the tire production process and working toward producing the highest quality product at the lowest possible cost are complementary objectives. Given that tire durability, rolling resistance, and high-speed performance are all very important parameters, and, in fact, that tires are certified as having met the relevant standards when the tire is shipped to the end-user, then uniformity is the greatest differentiator among many of the tire companies. This is dependent on the quality systems operating throughout the entire manufacturing process, including secondary variables such as tire component-to-component adhesion and hysteresis.

REFERENCES

1. Bhowmick AK, Hall MM, Benarey HA. *Rubber Products Manufacturing Technology*. CRC Press, Marcel Dekker, New York. 1994.
2. Posner J. Processability Tester. *Kautschuk Gummi Kunststoffe*, Vol. 56, No. 4, pp. 149–158. 2003.
3. Waddell W, Rodgers B. *Tire Applications of Elastomers 2. Casing*. Presented at a Meeting of The American Chemical Society Rubber Division, Grand Rapids, MI. 2004.
4. Tire uniformity. https://en.wikipedia.org/wiki/Tire_uniformity

10 Tire Testing and Performance

10.1 INTRODUCTION

Tire testing consists of three parts: i) laboratory testing; ii) general proving-ground testing, for which tires are mounted on vehicles that are then tested on specially prepared roads and test tracks, and iii) commercial fleet evaluations and market studies (1). Laboratory testing includes tire uniformity tests, speed rating determination, and assessment of rolling resistance, durability, and basic handling characteristics, such as cornering coefficient.

Proving grounds consist of high-speed test tracks, gravel roads, wet and dry skid pads, and tethered tracks for the testing of farm tractor tires. Testing is conducted under defined conditions, such as certain loads, inflation pressures, and speeds, and the use of specific vehicle types. The vehicles are specified to meet a set of defined conditions, such as wheel alignment, vehicle horsepower, wheelbase, and axle configuration.

Finally, commercial testing enables the tire engineer to obtain an assessment of how a design will perform under a broad range of service conditions that will be experienced by the end-user. The testing protocol which tires then undergo results in a broad range of products to meet the needs of both the vehicle manufacturers and the end-users for optimum performance under a variety of service conditions.

10.2 LABORATORY TESTING

Tire laboratory testing has undergone major revisions over the past number of years due to the regulatory authorities' desire to improve tire durability and reliability, reduce the environmental impact of tires, specifically rolling resistance and noise generation. For example, regarding noise, for a vehicle travelling at 80 kilometers per hour (kph) or 50 miles per hour (mph), up to 80% of the vehicle-generated noise can originate from the tires, and this can be amplified by the road surface. Laboratory tire testing falls into three broad categories:

1. Durability, including burst pressure, tire rim seating, and high speed
2. Rolling resistance and heat build-up
3. Inflation pressure loss rate and secondary tests such as noise evaluation

In many respects the governing standard methods covering these tests are the Federal Motor Vehicle Safety Standard FVMSS139 and the US Department of Transportation Uniform Tire Quality Grading (UTQG) procedure, to which further reference is recommended. These methods have then been adapted or modified by many governments and regulatory agencies to meet the needs of their specific markets; in many cases, the European standards for traction, noise, and rolling resistance have been the initial standard, which was then adapted by other agencies. In some instances, original equipment manufacturers require a plunger test and a bead-unseat test. These are also described in the test methods described later in this chapter. However, in most instances, endurance, rolling resistance, and high-speed capability remain the primary focus. Finally, the industry standard test methods are under continual review, updating, and improvement. As a result, some of the test conditions cited below may change, so it is always best to have current copies of the test protocols.

10.2.1 FMVSS139 Tests

Following regulations issued by the United States Congress, the test procedures used to assess a tire's durability, documented in FVMSS109, were updated and published as FMVSS139. The regulation is called Transportation Recall Enhancement, Accountability, and Documentation Act, or Tread Act, and, though it was very complex legislation, it had several key parts:

1. Define new tire endurance test requirements
2. Document and report all tire warranty claims and recalls. This is a global requirement for all manufacturers who sell tires in the United States.
3. Additional sidewall information
4. Vehicle tire inflation pressure monitoring capability

This review is not intended to discuss or review the details of the test methods described in FMVSS139, but rather to highlight the important points as they pertain to tire engineering and new product development. The endurance or durability test is conducted by installing a mounted and inflated tire on a dynamometer or against a flat-faced steel test wheel 1.7-m (67.23 inches) in diameter (Figure 10.1). For passenger car tires, this is a step-load test run as described below.

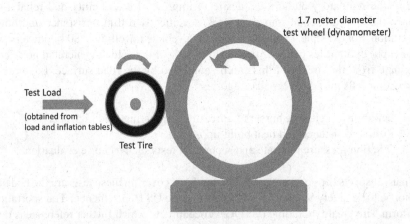

FIGURE 10.1   Schematic of 1.7-m Diameter Test Assembly

Though the test methods are described in detail in the standard FMVSS139 and the other national standards, a brief synopsis of each test method is appropriate.

10.2.1.1 Endurance

Room temperature	38°C
Speed	80 kph (50 mph)
Load	Set at the rated inflation from the Tire and Rim Association tables and also in the FMVSS139 specification document

1. 85% of rated load for 4 hours
2. 90% of rated load for 6 hours
3. 100% of rated for 24 hours

Pressure	Nominally 180 kpa (26 psi)

Tire companies may modify the test for internal research and development purposes, typically after completion of the 3rd step or after 34 hours of run time, by further increasing the load on the tire in 10% increments every 8 hours until tire failure or removal for other reasons.

10.2.1.2 Low Inflation Pressure Performance

On completion of the endurance test, a low-pressure performance test can follow. In this case, the same mounted tire that had been used for the endurance test on the 1.7-m diameter test wheel can be used again, assuming it had successfully passed the endurance test and there is no damage. The inflation pressure is set at 140 kpa (20.3 psi) and the test is then conducted as follows;

Room temperature	38°C
Speed	120 kph (75 mph)
Load	Set at 100% of the tire's maximum load-carrying capacity
Pressure	Nominally 140 kpa (160 kpa for heavy-load rates tires)
Duration	90 minutes

10.2.1.3 The High-Speed Test

The high-speed test is also conducted on a 1.7-m diameter test wheel or dynamometer. In this case, there is first a two-hour breaking-in period, where the tire is run under load for two hours at 80 kph or 50 mph, after which the tire is allowed to cool for up to another two hours. The test method is then as follows;

Room temperature	38°C
Speed	140 kph (87 mph) for 30 min
	150 kph (93 mph) for 30 min
	160 kph (99 mph) for 30 min
Load	Start at 85% of rated load
Pressure	Nominally 220 kpa (32 psi)

As in the case of the endurance test, upon completion of the test, tire companies may run extended testing by increasing the speed in 10-kph or 10-mph increments every eight hours, or for one shift, until tire failure. In many instances, the high-speed test can be a leading indicator of a tire construction's endurance performance, i.e., the better the high-speed performance, the better the tire durability and in-service performance. This often proves to be a valuable research tool in tire development work.

For extended durability testing of larger truck tires, a 3-m diameter test wheel, in place of the 1.7-m wheel, can be used. Because of the greater test wheel diameter resulting in less deflection of the tire crown region, the test will be run for longer but can allow a more in-depth identification of potential tire failure modes not evident by a more destructive test conducted on lower-diameter test wheels

10.2.1.4   Description of Potential Tire Removal Modes

Though there are many reasons for removing a tire from a test, the major factors are as follows:

1. Bead Separation: Breakdown of the components in the bead, component-to-component adhesion break-down, and separation
2. Belt Separation: A crack leading to a full separation, typically starting at the edge of the working belts, and then propagating between them. It can also start at the shoulder wedge interface or, in the case of heavy-duty truck tires, the 2/3 wedge (wedge located at the end of Belts #2 and #3).
3. Chunking: Tread or sidewall pieces breaking away from the tire
4. Cord Separation: Belt of ply cords separating from the adjacent rubber coating compounds
5. Cracking: Cracks and surface separation on the tread, sidewall, and external lower bead areas
6. Inner Liner Separation: Inner liner coming off the inside of the tire
7. Ply Separation: The ply end breaks loose and a crack emerges in the outside of the tire
8. Sidewall Separation: Sidewall component separates from the ply
9. Tread Separation: The tread compound separates from the upper belts or overlay.

10.2.2   Rolling Resistance

Tire rolling resistance is defined as the force or energy required to maintain the forward movement of a loaded pneumatic tire in a straight line at a constant speed. It is thus the scalar sum of all contact forces tangential to the test surface and parallel to the wheel plane of the tire. Tire rolling resistance is caused by the natural viscoelastic properties of rubber, along with the tire's internal components constantly bending, i.e., hysteresis, where the tread footprint under load flattens against the road surface and then recovers in the unloaded state as it exits the footprint. Aerodynamic drag, friction in the contact patch, and friction with the rim also contribute to the total rolling resistance, F_R, typically reported in [J/m] or simply [N]. Each of the standards organizations have written test methods to measure tire rolling, and some examples include:

- Society of Automotive Engineers, SAE J1269 - September 2006-09; Rolling Resistance Measurement Procedure for Passenger Car, Light Truck and Highway Truck and Bus Tires
- SAE J2452 - June 1999; Stepwise Coast-down Methodology for Measuring Tire Rolling Resistance
- International Standards Organization ISO 18164:2005(E); Passenger car, truck, bus, and motorcycle tires – contains methods of measuring rolling resistance, and is popular in Europe
- ISO28580; Passenger car, truck, and bus tires – methods of measuring rolling resistance – single-point test and correlation of measurement results

SAE J1269 has three different test methods, all run on the 1.7-m diameter test drum. These are i) force method, which measures force at the tire spindle, and converts it to rolling resistance, ii) torque method which measures the torque input at the machine, and iii) power method, which measures the power input to the test machine and again converts it to rolling resistance.

Rolling resistance of the free-rolling tire is the sum of the contact forces at a tangent to the test surface and parallel to the wheel plane of the tire, and the rolling resistance coefficient is the ratio of the rolling resistance to the load on the tire. Table 10.1 shows some examples of rolling resistance coefficients obtained by different techniques.

In addition to experimental methods for measuring rolling resistance, finite element analysis (FEA) is also used, and, in many instances, can be more accurate in terms of determining rolling losses due to whole-tire hysteresis. FEA allows the calculation of the contribution of each component in the tire to tire rolling resistance. Analyses of a wide variety of tires have been published and, in most instances, the distribution is dominated by the tread. The greater the traction requirement, the more hysteretic will be the tread and, in consequence, the energy losses will be the greatest. For a steer axle heavy-duty truck, Ford et al. published a representative model, summarized in Figure 10.2, showing that the tread region contributed 32% to whole-tire rolling resistance (2, 3). Other models have shown this value to be as high as 45%, which is still lower than that reported for

TABLE 10.1
Rolling Resistance Coefficient (RRC) Targets

Test Method	Example of Truck Tire Rolling Resistance Coefficients, RRC, (kg/metric ton)
SAE J 1269. 1.7-m test wheel	7.0
SAE J 1269 5-point Test, 1.7-m test wheel	7.0
ISO 28580, 2-m test wheel	6.6

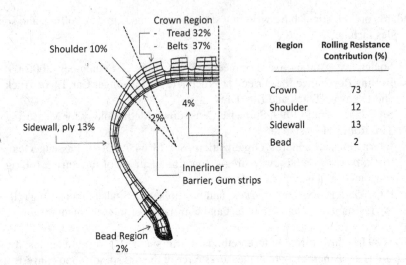

FIGURE 10.2 Finite Element Analysis of a Truck Tire and Component Contributions to Rolling Resistance (3)

passenger car tires. Typically, the larger the tread volume, such as in switching from a steer axle rib tire to a deep lug drive axle tire, the greater the tire rolling resistance.

The tread area thus represents a tire's largest contributor to rolling resistance and, in the case of a high-performance traction passenger tire, may account for up to 70% of the tire's rolling resistance. The estimated impact on vehicle fuel consumption of tire rolling resistance can vary considerably, but, as a guide, for passenger car tires, a 7 to 10% reduction in rolling resistance can achieve up to a 1% improvement in vehicle fuel economy. For truck tires, the impact is greater. A 2.5% to 3.5% reduction in rolling resistance can achieve up to a 1% reduction in fuel consumption in long-haul truck operations.

An example of other models is summarized in Figure 10.3 for both an all-season passenger car tire and a truck tire. In the case of high-performance passenger car tires, the contribution of the tread can be as high as 70% or even higher.

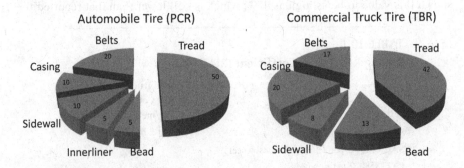

FIGURE 10.3 Rolling Resistance Contribution (4)

10.2.3 BEAD AREA DURABILITY

The tire bead area can affect tire durability by one of two major ways, either ply-end separation or other component-to-component separations, due to adhesion failures. The bead durability test can be conducted by one of two ways. In one method, the tread is buffed off, the tire is mounted and then run on a 1.7-m test wheel at 110% of its rated load. The load can then be increased in 10% increments either every eight hours or every 1,000 km increments until the tire fails. Alternatively, if the test wheel is designed appropriately, the tire can be mounted inside the wheel, and a similar test sequence is followed. In both cases, the objective is to transfer the maximum stress to the bead region rather than the crown area (Figure 10.4).

10.2.4 INFLATION PRESSURE RETENTION

Tire inflation pressure retention or inflation pressure loss rate (IPLR) is a significant differentiator between high-quality and medium-quality tires. Accurate measurement requires rigorous control of test temperatures and monitoring of atmospheric pressure to enable standardized loss rate calculations. The test method for passenger tires (PCR) is briefly:

1. Temperature is controlled to $21 \pm 0.3°C$
2. Two tires are mounted on rims, and conditioned in the temperature-controlled laboratory until an equilibrium is achieved
3. Testing starts with the tires inflated to 2.4 Bar
4. Inflation pressure is then monitored for a defined period, ranging from 42 days onward to enable an inflation pressure loss rate to be calculated

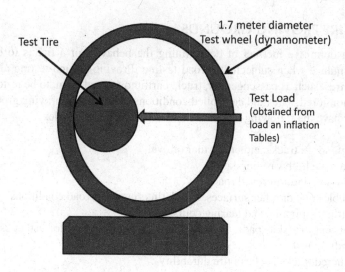

FIGURE 10.4   Bead Durability

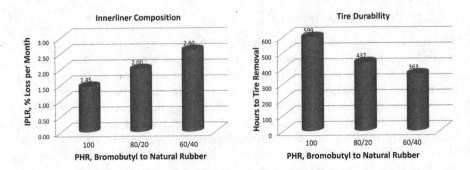

FIGURE 10.5   Inner Liner Composition and Tire Durability (5)

The *de facto* industry standard for passenger car tires is a maximum inflation pressure loss rate of 2.5% per month, and this was the standard set by General Motors, followed by other original equipment manufacturers. Truck tires, with inner liner gauges of 2.0 mm and containing 100 phr of bromobutyl, have inflation pressure loss rates as low as 0.5% per month. Such low inflation pressure loss rates are believed to be important in ensuring low adjustments or warranty claims, particularly for heavy-duty commercial truck tires.

A secondary test conducted many times with IPLR is intra-carcass pressure (ICP). This is conducted by inserting pressure gauges on the sidewall, which have needles that penetrate to the ply cord. When properly installed, gas pressure due to air permeating through the tire casing can be measured. The lower the permeability of the inner liner, the lower will be the intra-carcass pressure, with a lower oxygen build-up within the body of the tire. This results in lower heat history and, in turn, results in less oxidative degradation of internal components over time allowing better tire durability (Figure 10.5).

10.3   TIRE PROVING GROUNDS

The most definitive method of determining the behavior of a tire is to examine its performance when subjected to road testing. Proving-ground testing allows all types of tires, such as passenger car, truck, earthmover, and farm, to be tested under closely monitored, safe, and controlled conditions. An industry proving ground will generally have the following test tracks and road courses available:

1. High-speed tracks, either circular or oval
2. Interstate highway simulation
3. Gravel and unimproved roads
4. Cobblestone or other surfaces, simulating rough off-road conditions
5. Cutting, chipping, and tearing courses
6. Wet and dry skid pads, serpentine and slalom courses for esthetics, and handling tests
7. Tethered tracks for farm tire durability
8. Glass roads for footprint monitoring

The area of the proving ground is, in many instances, over 30 square kilometers and will have service bays, garages, and laboratories to meet the testing requirements (Figure 10.6).

10.3.1 TRACTION TESTING

The longitudinal force is that force acting in the direction of or opposite to the direction of travel and is generated by acceleration or braking. Tractive forces depend on two primary factors, the tire tread (both design and compound), and the road surface.

Test facilities would have multiple tracks for testing traction but would have, as a minimum, three surfaces, namely concrete, macadam or asphalt, and a surface with a low coefficient of friction to mimic ice, such as might be obtained with a blue basalt surface.

Road surfaces can create the largest variance in traction. Driving on wet, dry, ice, snow, or mud are just some examples that necessitate the design engineer to optimize the tread for the widest range of environmental conditions drivers will experience. Braking causes the wheel to slow, in turn causing longitudinal slip. Longitudinal slip is essentially zero for a free rolling tire but, in practice, is 100% at lockup. It is expressed as a percent as shown in Equation 10.1.

$$\% \text{slip} = \Big[\big(\text{Vehicle Speed} - \text{Tire Speed}\big) / \text{Vehicle Speed}\Big] \times 100 \qquad (10.1)$$

A simple graphical representation of the traction coefficient is shown in Figure 10.6. The peak braking traction is the maximum force developed with wheel lockup and the peak traction coefficient is the ratio of the peak traction force to the vertical load.

The slide value is developed when the wheel is locked and is the ratio of slide traction coefficient ($Fx_{.slide}/Fz$) to the vertical load (Figure 10.7). Tire load changes can affect the traction coefficient on both wet and dry surfaces. Similarly, for high speeds on wet surfaces, hydroplaning will also contribute to traction losses.

Hydroplaning is the phenomenon of reduced tire traction performance on wet surfaces and is particularly evident at high speeds. Hydroplaning is the lifting of

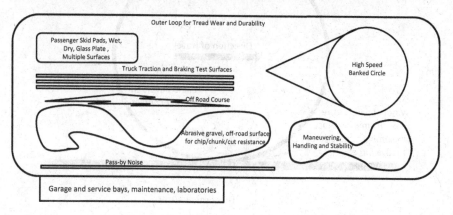

FIGURE 10.6 Idealized Tire Proving Ground

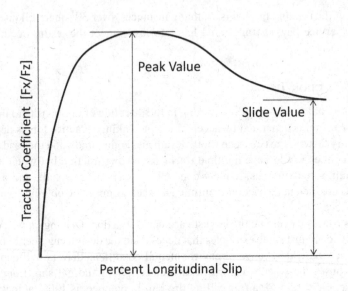

FIGURE 10.7   Traction Coefficient versus Percent Slip (2)

the tire tread surface off the road due to the layer of water. Total hydroplaning is the complete separation of the tread from the road surface, i.e.,

$$\% \text{Hydroplaning} = \left[\text{Dynamic Footprint Area} / \text{Static Footprint Area} \right] \times 100 \quad (10.2)$$

In hydroplaning, three zones in the tire-to-road contact area have been identified (Figure 10.8):

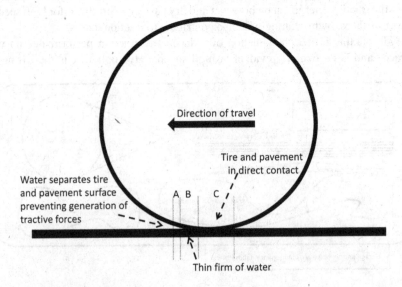

FIGURE 10.8   Hydroplaning Zones

- A: the front or leading edge of the footprint, where water hydrodynamic pressure is sufficient to prevent the tread from making contact with the pavement
- B: where most of the water has been expelled but a thin film remains, maintaining partial separation of the tread from the pavement
- C: full tread–pavement contact, and tractive and frictional forces are generated

The three zones are visually evident using glass plates, where a vehicle is driven at speeds up to 120 kph over a glass plate located above a high-speed camera. A film of water, nominally 2.0 mm in depth, can be sufficient to create the three zones in hydroplaning.

Therefore, in summary, three factors will impact hydroplaning and, in turn, tire tractive qualities, namely water conditions and depth on the road surface, pressure distribution across the tire-to-road contact patch, and the capacity of the tire tread pattern to disperse the water film on the road surface. These factors are of considerable importance when designing tread patterns, groove depths, and tire crown constructions to ensure optimum footprint pressures.

10.3.2 Noise

Up to 80% of the noise generated by a vehicle travelling at 80 kph is generated by the tires, and this is why noise has been incorporated into many regulatory standards, such as the European Labelling requirement. Regarding the tire, there are a number of contributory structural parameters responsible for noise generation;

- Tread design, pitch, and sequencing. Air is trapped and compressed between the tread and the pavement. As high-pressure air is then released at high frequencies, noise is emitted
- Lug aspect ratio, stick-slip, and snap-back. As the loaded tread elements roll onto and then out of the pavement contact area, high stresses are released. The resulting vibrations, analogous to vibrations from a tuning fork, are frequently the major component of tire noise
- Sidewall deflection and resonance
- Radial and tangential vibrations
- Air cavity resonance

Sidewall deflection, and radial and tangential vibrations are best studied in sound chambers. By running the mounted tires on a smooth drum, the effect of the pavement can be eliminated, thereby allowing measurement of frequencies attributable only to the tire casing. Inflation pressure can affect air cavity resonance and cavitation amplification, which is why addition of a damping foam inside the tire is being studied by the industry.

In addition to tire structural parameters, operational factors, such as speed, vehicle load, and pavement surface, can also affect noise generation. Concrete would tend to generate the highest tire–pavement noise (78 to 80 dB(A)) while newly laid asphalt

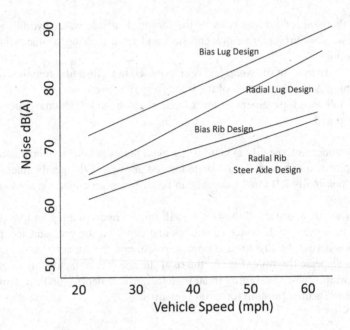

FIGURE 10.9 Radial *versus* Bias Truck Tire Speed *versus* Noise Generation (1, 2, 3)

would be the lowest (70–72 dB(A)). Tire noise generated by a vehicle driving under steady-state conditions can be measured on test tracks where the vehicle passes pre-positioned microphones as described in SAE J57. Figure 10.9 shows the relative noise generated by various truck tires and the effect of speed. Lug tread designs would tend to show the greatest increase in noise with increasing speed (2) (Figure 10.10).

In addition to test track measurements and laboratory sound chambers, finite element analysis is now proving to be a valuable tool in this area and offers the best opportunity to further understand this phenomenon.

10.3.3 TREAD WEAR

The wear of a tire with service is a very complex phenomenon and, to date, there is no clear understanding of the fundamental mechanism by which this occurs. This is due primarily to the wide range of service conditions and environments under which vehicles operate. A slipping action between the tire and the road surface generates friction, which is thus dependent on intensity and duration, i.e., load, speed, and time. Given this, wear could be attributed to one of two mechanisms, either i) thermo-oxidative degradation of the tread compound, which would occur under slow-wear conditions, or ii) a tensile-tearing mechanism, whereby rubber is torn from the tread. The Schallamach wave theory would fit under this description and would occur under faster-wearing conditions (6, 7).

In practice, wear can be described in terms of abrasion rate:

1. On-off road, where tread tearing, chipping, and chunking occur, reflecting fast-wear conditions

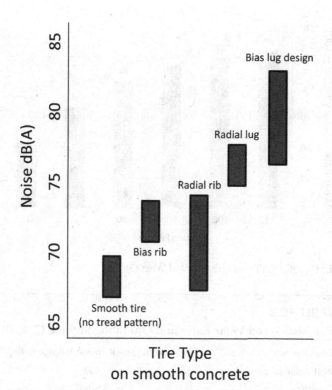

FIGURE 10.10 Tire Construction and Tread Design on Noise Generation (2)

2. Short-haul stop-start conditions on paved roads as would be found for city bus, city taxi, and short commutes, representing again, fast-wear conditions
3. Medium-haul routes and commutes
4. Long-haul routes, such as would be seen for trucks hauling loads on interstate highways, which would be considered slow-wear conditions.

This is of significance in designing test tracks to mimic various road conditions. A proving- ground facility would therefore have multiple tracks representing on-off road conditions, short-haul abrasive surfaces, and then long-haul slow-wearing conditions, which may extend into the use of public highways under controlled conditions. In practice, the primary operational variables impacting wear will be inflation pressure (Figure 10.11), alignment, load, speed, and environmental conditions, such as temperature and topography. Table 10.2 shows a compilation of potential wear rates on different road surfaces.

Seasonal conditions are also important (Figure 2.18). Speed and wear exhibit a near-linear positive relationship, with faster speeds creating faster wear rates for all tire types.

In on-off road services, the wear mechanism would be predominantly a cutting-and-chipping process. A tread compound with a high tensile and tear strength, therefore, becomes of importance, so such tread compounds tend to be predominantly

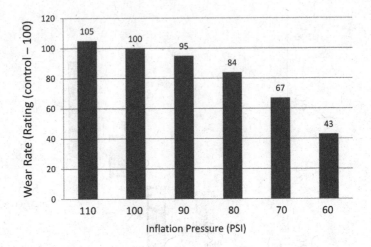

FIGURE 10.11   Inflation Pressure and Tread Wear (3)

TABLE 10.2
Potential Tread Wear Rates and Operating Surface (3, 8, 9)

Road Surface	Potential Tire Removal Mileage Rating
Well-maintained asphalt	100
Concrete	95–100
Macadam (medium)	90–100
Macadam (coarse)	80–90
Gravel	40–65
Off-road (dirt and rock surface)	20–40

composed of natural rubber, compound filler or reinforcement systems designed to improve tear strength, with tensile strengths greater than 24 MPa and high elongations at break.

Under such conditions, low inflation pressure will help reduce road contact forces, thereby reducing the tendency for tread cutting and tearing, and thus improving tread durability. Bias tires may also demonstrate this effect.

10.3.4   TIRE VEHICLE HANDLING

High-speed tracks at the proving ground can allow both subjective evaluation and quantitative assessments of the handling characteristics of tires. Subjective evaluation would be determined by skilled test drivers and included assessments such as:

1. Steering precision
2. Road-holding
3. Comfort

4. Aligning torque
5. Noise
6. Vibration
7. Traction and grip
8. Stopping distance
9. Cornering

Vehicle alignment will have an effect on tire–vehicle handling characteristics. In the case of trucks, there are three alignment settings of importance, namely camber, caster, and toe. Camber is the tilt of the front wheels on a vehicle (Figure 10.12). Outward tilt from the vertical axis Z, is positive camber, with inward tilt at the top being negative camber. Toe describes the orientation of the tires, i.e., pointing outward, which would be a toe-out condition, or inward, which is a toe-in condition (Figure 10.13). Vehicle alignment will be set by the manufacturer, and the owner would ensure that the settings are maintained by regular maintenance programs.

In terms of vehicle wheel alignment, two other terms are important, toe and caster. Caster refers to the setting of the kingpin which enables front wheel steering (Figure 10.14). Caster is set by the kingpin and is therefore not adjustable. Camber and toe settings, which are typically set when the vehicle is built, are adjustable and can be reset to factory conditions when new tires are installed. When the correct alignment has been verified, handling and maneuverability testing on a track can then be conducted.

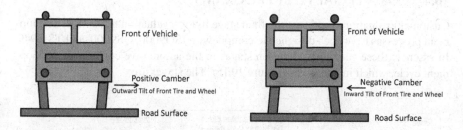

FIGURE 10.12 Alignment: Definition of Camber

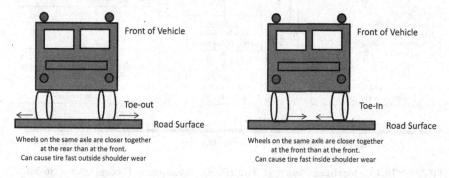

FIGURE 10.13 Alignment: Definition of Toe-In and Toe-Out

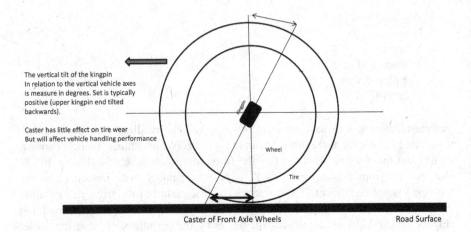

FIGURE 10.14 Kingpin Setting

In conclusion, other than coast-down tests, proving grounds do not necessarily lend themselves to tire–vehicle rolling resistance testing and fuel consumption evaluations. After the tire has been tested for rolling resistance on the 1.7-m diameter test wheels, and a coast-down test has been conducted, tire–vehicle fuel consumption testing is best conducted under controlled conditions on highways.

10.4 COMMERCIAL FLEET PROGRAMS

Customer evaluations represent the final stage in tire evaluation tests. Most development processes have been designed to comply with ISO9000, QS9000, and TS16949. In essence, these standards set seven stages in the automotive component development cycle, which include tires (Figure 10.15). The stages are:

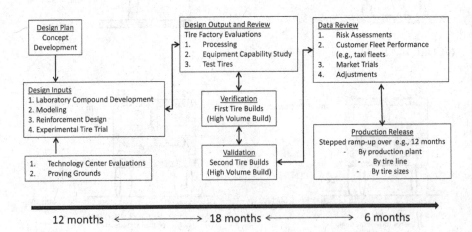

FIGURE 10.15 Idealized Passenger Tire (PCR) Development Process over a 36-Month Period

1. Design plan, where performance targets are defined
2. Design inputs, which include laboratory data and initial design parameters
3. Design output, including initial development modelling, materials, and construction data
4. Verification, including initial prototype tires
5. Validation of test data, including production trials, risk assessments, and customer feedback
6. Final data review
7. Product release.

In the final steps, customer and fleet data are of importance in ensuring a trouble-free product release. In the case of truck tires, fleet testing is the final step before product release. Testing would typically collect the following information: tread wear rate, resistance to irregular wear, handling, and fuel economy. Fuel consumption has emerged as one of the most important performance parameters, due to the growing consumer awareness of the environmental impact of emissions. The development time for most companies following this protocol is around 3 years. For commercial tires, such as for heavy-duty trucks and farming, the development time can extend up to five years and, in the case of aircraft, potentially longer.

10.4.1 Truck Fuel Consumption

In the case of commercial truck tires, there has been a significant improvement in rolling resistance over the past 30 years (Figure 10.16). Many factors have come together to achieve this improvement, but the major factors were radialization, which is still continuing in many regions of the world, and low-profile tire constructions.

There are five factors influencing truck fuel consumption:

1. Aerodynamic losses
2. Axle alignment
3. Drive train losses
4. Accessories, such as heaters and compressors
5. Tires

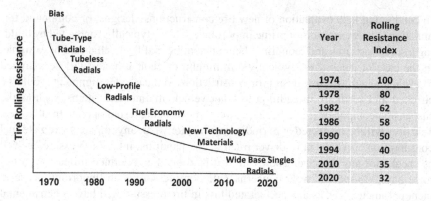

FIGURE 10.16 Truck Tire Rolling Resistance Improvement (6, 7)

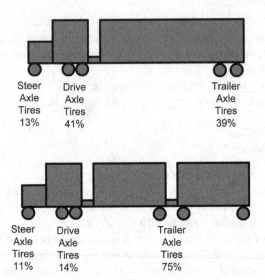

FIGURE 10.17   Truck Tire Rolling Resistance Contribution by Axle Position

For a tractor-trailer combination travelling at 88 kph (55 mph), 42% of the horsepower is required to overcome aerodynamic drag, 15% for drive train losses, 9% for accessories, and 34% for tire rolling resistance (5). The relative losses change with vehicle speed and load, but tires remain a major contributor to energy consumption of a highway truck.

Of the energy expended in overcoming tire rolling resistance, this can be further dissected, showing that the steer axle accounts for 13 to14% of the vehicle tire system rolling resistance losses, the drive axle 41%, and the trailer axle 39% (Figure 10.17).

The conversion to radial tires has had the largest impact on fuel savings and is why the trend toward radialization continues (Figure 10.18). Substantial savings are possible by installing and maintaining energy-efficient low-rolling-resistance tires.

10.4.2   COMMERCIAL FLEET TIRE WEAR

In conducting field evaluation of new tire constructions, designs, or compound formulations, the wear test is of prime importance. This is typically conducted by building a set of test tires and mounting them on commercial fleet vehicles. For example, in the case of truck tires, typically a minimum of eight vehicles will be assigned to evaluate each tire test design or construction. Vehicle axle alignment would be checked and certified, in addition to other vehicle maintenance items, such as the suspension components, to ensure that the vehicles being tested operate trouble free. Test tires will be inspected at periodic intervals for wear, any impact on vehicle fuel consumption, assessment of driver traction and handling, inflation pressure, as well as checks for any damage. They will also be rotated at a defined mileage to minimize any axle-specific wear concerns. Inflation pressure is an important measurement parameter (11), as any accelerated loss in tire pressure will have a detrimental impact on wear resistance (Figure 10.19). Figure 10.20 illustrates the impact of axle

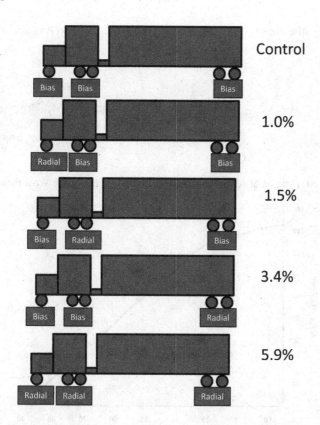

FIGURE 10.18 Truck Fuel Savings with Conversion to Radial Tires (2, 6)

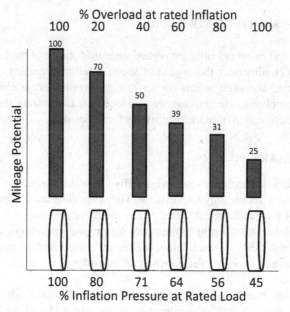

FIGURE 10.19 Inflation Pressure and Tire Wear Potential (11)

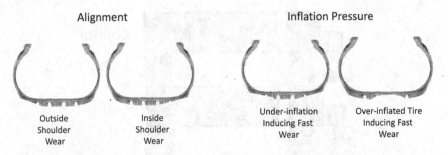

FIGURE 10.20   Impact of Alignment and Inflation Pressure on Tire Wear Patterns (11)

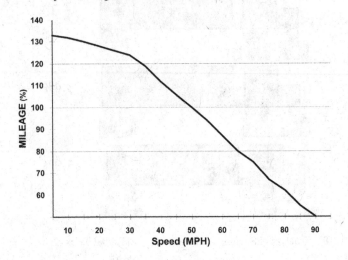

FIGURE 10.21   Speed and Tread Wear (11)

misalignment and improper inflation on tire wear uniformity of passenger car tires, and Figure 10.21 illustrates the impact of speed. The consequences of inadequate inflation pressure and high speeds on performance results occur for all tire lines. These points emphasize the importance of adequately controlling the field evaluations to ensure collection of representative performance data.

10.5   SUMMARY

A manufacturer's tire testing resources typically fall under the aegis of the Research, Development, and Technology Department. They provide a service role but are still independent of the development teams. There are three divisions within the testing organization: laboratory testing group, proving-ground operations, and engineers working with test fleets. For the latter, the engineering test fleet team would also provide a technical service role, which is of considerable value to the Sales and Marketing group.

In tire testing, tread wear is one area that shows how the load, in conjunction with inflation pressure, can determine tire performance. Tread wear is adversely affected by incorrect or heavy over-loading, which results in excessive deflection of the tire. Overloading or under-inflation creates fast wear regions on the shoulder of the tire. Likewise, an inflation pressure too high for the load could cause rapid wear in the center of the tire. Inflation and loading conditions also affect handling, steering response, fuel economy, and durability of the product.

REFERENCES

1. Kovac F. *Tire Technology.* The Goodyear Tire & Rubber Company, Akron, OH. 1978.
2. Ford, TL, Charles FS. *Heavy Duty Truck Tire Engineering.* Society of Automotive Engineers, SP729. Warrendale, PA. 1988.
3. Mezynski SM, Rodgers B. *Radial Medium Truck Tire Performance and Materials.* American Chemical Society Rubber Division, Akron Rubber Group, Akron, OH. 1989.
4. Waddell W, Rodgers B. *Tire Applications of Elastomers, 2 Casing.* Presented at a Meeting of the American Chemical Society, Rubber Division, Grand Rapids, MI. 2004.
5. Waddell W, Rodgers B. *Tire Applications of Elastomers, 3 Innerliner.* Presented at a Meeting of the American Chemical Society, Rubber Division, Grand Rapids, MI. 2004.
6. Mezynski SM, Rodgers B. *Radial Medium Truck Tire Performance and Materials.* Presented at a Meeting of the American Chemical Society Rubber Division, Louisville, KY. 1990.
7. Bridgestone, www.commercial.bridgestone.com/en-us/truck-and-bus.com.
8. Rodgers MB. ELL Technologies. *Tire Wear Performance Estimates.* Cedar Park, TX. 2020.
9. Smith RH. *Analyzing Friction in the Design of Rubber Products and Their Paired Surfaces.* CRC Press, Boca Raton, FL. 2007.
10. Muhr AH, Roberts AD. Friction and Wear. Page 773-815. In *Natural Rubber Science and Technology.* Ed. AD Roberts. Oxford Science Publications, New York. 1988.
11. Gabor J, Wall J, Rodgers B. *Overview of Medium Radial Truck and Off the Road Tire Technology.* Presented at a Meeting of the American Chemical Society Rubber Division, Providence, RI. 2001.
12. SAE J57. *Sound Level of Highway Truck Tires.* Society of Automotive Engineers, Detroit, MI. 2005.

11 Future Trends

11.1 INTRODUCTION

Many of the tire industry predictions and trends that have been published over the past twenty years have not materialized. There are several reasons for this, such as costs in redesigning the vehicle corner, inadequate product durability performance, and manufacturing complexities (Figure 11.1). Conversely, other disruptive technologies have been successful and have resulted in many companies losing the ability to compete. Examples are radialization, elimination of the tire inner-tube, the "green" tire, silica tread technology, increasing the diameter of low aspect ratio tires, and manufacturing automation. In many instances the cost of responding to such innovations and retrofitting manufacturing plants tires has been prohibitive, requiring construction of new "green-field" plants.

As of 2018, the global automotive tubeless tire market is expected to grow at a rate of up to 6% per year and approach a value of up to $196 billion by the end of 2024. Most of this industry growth will be due to a variety of reasons: i) emerging markets in developing countries and the increasing number of passenger cars, ii) government regulations, and iii) technological advancements in tubeless tires replacing tube-type tires in severe or off-road service applications.

The passenger car segment is expected to account for the largest share of tire company revenue across the globe. However, for a tire company to be truly successful, there is the hypothesis that a successful heavy-duty truck tire is essential, i.e., passenger tires create the volume but commercial radial truck tires generate the profits, though this conclusion also depends on the market location. Rising sales of passenger car vehicles across the globe is accelerating the growth of the automotive tubeless tire market. The growth of the tubeless radial tire sector can also be attributed to early damage to the tires caused by poor and inadequate road infrastructures. For instance, in South East Asia, less than 40% of roads in use are developed and modernized. Substandard road infrastructure can result in premature damage to the tires, which is raising the demand for tires in the replacement market. Thus, growing demand and sale of passenger car vehicles in the world is expected to strengthen the growth of the automotive tubeless tires along with increasing preferences expressed by consumers for tubeless tires over tires with tubes is expected to spur the growth of the market. To react to such evolutionary and revolutionary changes in tire technology, three issues are pertinent:

1. Consideration of technological innovations (projected and being implemented) and potential changes in the regulatory environment
2. Original equipment innovations, such as vehicle power train changes
3. Competitive threat analysis to minimize the risk of being confronted with disruptive industry change.

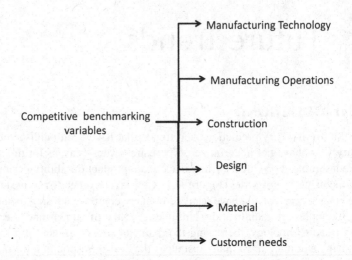

FIGURE 11.1   Elements in Future Trends and Competitive Technical Intelligence

New innovations fall under each of the core disciplines of tire technology, i.e., materials science, construction mechanics, and design, with materials science, in this case, being the critical enabler. New materials being investigated that have a significant potential for market adoption in the next ten years have been summarized in Table 11.1.

Of the new materials technology trends, functionalized solution SBR is the most promising in that it allows an increase in the reinforcement potential, improved polymer–silica interaction and reduction of the Payne effect, implying better abrasion resistance and lower hysteresis. Though the cost increase for the polymer may be up to $500 per ton, it would still be beneficial in some premium tire lines in achieving the conflicting performance needs of traction, rolling resistance, and abrasion resistance (1, 2).

Looking forward over the next ten years, several trends may have an impact on materials technologies. First, there has been a shift in the cost breakdown of a tire (2). Raw materials costs have increased substantially and these increases are expected to continue. As automation has increased, driven largely by the shortage of skilled labor, associated costs have decreased (Table 11.2), and with greater plant efficiencies, conversion costs and other overheads have decreased.

Finally, in several parts of the world there has been a considerable amount of new natural gas production capacity coming on-line, particularly in the United States and in the Middle East region, in countries such as Qatar. Low-cost C_2 streams (ethane), readily convertible to ethylene and other unsaturated molecules, and C_3 (propane), which is convertible to propylene, offer new sources of low-cost feedstock from which new polymers can be developed. Ethylene is easily dimerized to butadiene and processed from there (Figure 6.7, 6.8), or new polymer derivatives could be developed using propylene and other new low-cost feedstocks.

TABLE 11.1
New Materials Technologies

Material	Application	Potential Benefit
New-generation SSBR	Tread compounding	Lower the Payne Effect, abrasion improvement, lower hysteresis
Natural rubber	All compounds	On-going radialization
Polyurethanes	All components	Non-pneumatic tires
Novel polymers	All compounds	Low-cost *versus* NR, SBR, BR
Bio-sourced monomers	Synthetic rubber	BR alternative
Bio-sourced process oils	Lubricants	Environmental
Hybrid fillers (carbon black)	Tread compounds	Abrasion and hysteresis
New silicas (HDS)	Tread compounds	Rolling resistance, wear
Recycling (post-industrial)	All compounds	Environmental
Re-use	Retreading	Environmental
Hybrid cords	Reinforcements	Lightweight
Next-generation steel cord	Reinforcement	Lightweight
Reversion resistors	Vulcanization system	Curing, durability
Nanocomposites	Inner liner	Air retention
Films	Inner liner	Air retention, durability

TABLE 11.2
Elements of Tire Cost

Year	Raw Materials (%)	Labor (%)	Overheads and Conversion (%)
1985	50	25	25
2000	60	20	20
2015	70	15	15
2025	80	10	10

11.2 PASSENGER CAR TIRE TRENDS

There are a variety of trends occurring in the automobile tire market. These include regulatory labeling, environmental footprint reduction, vehicle tire comfort, dependability, and rolling resistance. Figure 11.2 shows the European label required on tires sold in that market and focuses on rolling resistance, traction, and noise generation.

It is expected that regulatory requirements will continue to drive the passenger tire market, followed by the needs of the original equipment manufacturers. Furthermore, other than the United States, most regulatory agencies are basing their labeling guidelines on the European standard. In the US, there may be more stringent requirements. For example, with regard to rolling resistance, although the test is very important, it only provides a "snap-shot in time", and does not offer a true representation of the fuel consumption contribution attributable to the tires on a

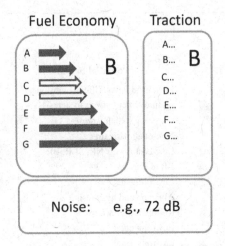

Fuel Economy   Traction

Fuel Efficiency
Based on rolling resistance measurements.
Rating "A" is best. Improving from "G" to "A"
Potentially allows up to 7.5% fuel savings
over life of tire

Traction
"A" rated tires perform best on wet roads
Determined by stopping resistance on wet
Surfaces, 80 km/h to zero.

Noise: e.g., 72 dB

Noise
External pass-by noise, 80 km/h

FIGURE 11.2 Elements of Regulatory Labeling. Core Elements of European Requirements Are Being Adapted by Other National Regulatory Bodies

vehicle. In-service vehicle fuel consumption due to tire rolling resistance also needs to consider changes in rolling resistance as the tread depth wears down, and the shift in whole-tire hysteresis with dynamic aging and air pressure loss rates (IPLR). Comfort, tire–vehicle stability, and durability (safety) are also critical parameters in the US operating environment. It is expected that these variables will eventually be captured in future labeling and regulatory guidelines.

However, with further regard to fuel efficiency, it is expected that, though rolling resistance will be a key element in meeting vehicle energy efficiency targets, the environmental impact of the tire will extend beyond CO_2 emissions to manufacturing waste, sometimes referred to as post-industrial waste, as distinct from post-consumer waste, manufacturing pollution (VOCs), and end-of-life disposal. Biomaterials will be a key element in minimizing the environmental footprint of tires (3).

In any case, the regulatory push to improve fuel efficiency will continue, again driven by the need to reduce vehicle pollution, reduce CO_2 and other tail pipe emissions, and reduce oil imports for those economies dependent on it.

The automobile tire industry could be considered to fall into several segments such as,

- Baseline tires
- Family sedan, all-season broad market
- Winter
- High-performance
- Ultra-high performance
- Radial light truck
- In the future, special-purpose tires for electric, autonomous, and shared vehicles

In terms of construction, tire diameters will continue to increase while aspect ratios will decrease, thereby allowing on-going improvements in fuel efficiency, and tire–vehicle handling and stability. The market will continue to segment, with the emergence of eco-tires in the broader market rather than a niche line, this being facilitated by biomaterials technologies.

Run-flat technology has been another new technology. Run-flat tires are based on one of two sets of technologies, the first based on an internal support mounted on customized rims, as developed by Michelin, and the second based on heavy-gauge, stiff, reinforced sidewalls, as developed by The Goodyear Tire & Rubber Company, and which can use existing rims. For both technologies, in the event of tire air pressure loss, either the internal support mounted on the rim or the stiff reinforced sidewall will maintain the tire's load-bearing ability until the vehicle can be driven to a location for repair (Figure 11.3). In-service such tires mounted on a vehicle travelling at a nominal speed of 80kph will rotate at a frequency of the order of 10 to 15 Hz. Under load, the sidewalls will flex with each rotation resulting in hysteresis heat build-up. The greater the sidewall gauge, and stiffness, the greater the heat generation will be and this has had a potentially detrimental impact on run-flat tire endurance.

Non-pneumatic tires have attracted considerable attention in the media and will continue to do so (Figure 11.4). Such assemblies have several advantages:

1. They offer improved mobility in military applications
2. They offer improved off-road capability on construction sites, such as for skid-steers
3. They offer potential to reduce manufacturing complexity since injection molding operations could be adapted
4. This technology promotes the tire manufacturer as an innovator and can generate very favorable publicity in the automotive media.

Many manufacturers have developed non-pneumatic constructions, but, in many respects, these can all fall into one of three configurations, namely hexagonal spring matrix, vertical or semi-vertical struts, or a coil configuration mimicking a spring mechanism. The support structures are produced with polyurethane. These structures connect to an inner steel hub and outside to a rigid rim with an outer rubber tread compound. This therefore provides the appropriate degree of both lateral and vertical stiffness.

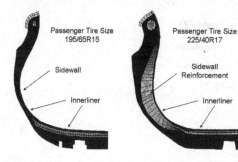

Run-flat Tires

- Improved mobility in run-flat conditions.
- Increased safety.
- Driving stability.
- Eliminate the need of spare tire.

FIGURE 11.3   Reinforced Sidewall Run-Flat Tire (4)

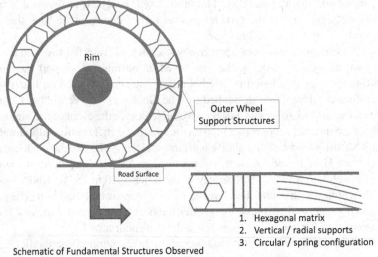

Schematic of Fundamental Structures Observed
in Prototype Non-pneumatic Wheel Assemblies

FIGURE 11.4 Non-Pneumatic Wheel Assembly Schematics

Challenges remain, such as tire–vehicle ride and handling, comfort, noise generation, flat spotting, and the durability of polyurethanes used in the assemblies. To address the tire–vehicle handling characteristics, redesign of the vehicle corner, to improve damping properties and compensate for the change in wheel spring rate, is needed, thus limiting the near- and medium-term potential application to original equipment (OE) manufacturers, and affecting potential retrofits. In consequence, given the existing global manufacturing infrastructure, the durable and reliable performance of the current radial tires with a 100 phr bromobutyl inner liner compound, and the challenges of vehicle corner redesign, more research could be needed to successfully displace the current technology.

It is expected that bio-sourced raw materials will find greater application and many concept tires, developed and displayed by tire companies, have demonstrated their utility. The range of materials is quite broad but a simple snapshot of what could be available is listed in Table 11.3 (2). Of these materials, guayule as a natural rubber

TABLE 11.3
Potential Bio-Materials for Tires (2)

Material	Application
Guayule	Replace natural rubber obtained from *Hevea brasiliensis*
Rice husks	Replacement for silica
Starch	Filler
Pine tar	Tackifying resin
Rosin	Rosin acid tackifying resin
Rice bran oil	Processing aid
Vitamin E	Antioxidant

source, rice husks as a silica source, and pine tar as a process resin have considerable potential for growth.

The use of electric vehicles (EVs) is on the rise and a potential impact of electric powertrains is the increased complexity of tire varieties. This includes construction optimization, such as aspect ratio and footprint changes, sizes, and proliferation of further SKU (stock keeping units) as a result of increased variation in OE tire types. Tire wear concerns with EVs make higher wear resistance critical, since traditional tires wear 30% faster on EVs than on conventional vehicles. Many EV tire changes also apply to autonomous vehicles (likely to be all or mostly electric), but the introduction and spread of autonomous driving means that further changes are emerging and will have to be scaled up alongside more traditional manufacturing. Emphasis on low noise and high-ride quality tires will increase. Reliability requirements may be higher.

Electric and autonomous vehicles are going through an evolution and are sometimes referred to as Connected, Autonomous, Shared and Electric (CASE). The Society of Automotive Engineers has defined six levels of automation (Table 11.4). Within each class of vehicle, namely sub-compact, compact, mid-size, sports, and full-size luxury, autonomous vehicles would expect to show weight increases due to the battery, power train, and associated electrical equipment. The increasing weight will require tires with higher load ratings. Inflation pressure loss rate would also become a critical performance factor, particularly in the "Shared" and "Autonomous" segments, since the decrease in driver engagement compared to that for "zero-automation" vehicles could impact routine maintenance activities.

Though the market is still developing, it therefore could be envisaged that, over the next 10 years, the automobile tire industry would follow a CASE track, while the heavy-duty light truck and commercial tire market would stay with the more traditional powertrains, based on the internal combustion engine, albeit with greater energy efficiencies. In any case, inflation pressure retention, or IPR, and reliability will be critical performance parameters given their impact on vehicle energy consumption and vehicle comfort.

There is still some uncertainly as to how new drive train systems on CASE platforms will impact tire design, with media reports describing quite diverse options

TABLE 11.4

SAE Classification of Connected, Autonomous, Shared and Electric (CASE) Vehicles (5, 6)

SAE Level	Control	Conditions Monitoring
0	No automation	Human driver
1	Driver assistance	Human driver
2	Partial automation	Human driver
3	Conditional automation	System but with human intervention
4	High automation	System
5	Full automation	System: all driving modes

TABLE 11.5

Vehicle Drive System Variable Which May Impact Tire Design

Drive Train Energy Sources	Challenges Operations	Environmental Impact
1. Battery electric	1. Unproven reliability	1. Fuel energy content
2. Fuel cell hybrid electric	2. Unproven maintenance	2. Energy conversion efficiency
3. Compressed natural gas	3. Effective service life	3. Vehicle design
4. Liquefied natural gas	4. Vehicle infrastructure support	4. Tare weight, payload capability
5. Propane	5. All region, all season capability	5. Freight-ton efficiency (TMPH)
6. Hybrid diesel electric	6. Operational insurance costs	6. Gross energy conversion efficiency
7. Biodiesel	7. Retention/residual value	7. Energy costs
8. Biogasoline		8. Fuel source infrastructure
		9. Fill-time
		10. Drive train system costs
		11. Maintenance
		12. Secondary pollutants

that might evolve. Table 11.5 lists current drive train systems that may come into production in the next 15 years, the operational challenges such systems might present, and then potential variables that may have both positive and detrimental effects on the success of such technologies.

Finally, the importance of developing "concept" tires and using Formula 1 and Indy racing tires for concept evaluation cannot be emphasized enough. Many new tire technologies, such as resin technology and carbon fibers, have been studied first on the race track, then optimized, and then successfully translated into general production.

11.3 COMMERCIAL TRUCK TIRES

There are two aspects to future trends in commercial truck tires, i.e., those tires used on highway trucks. First, there is the on-going downsizing of tires. For example, in the 1980's, the 24.5-inch diameter tire was dominant. Sizes such as 12R24.5 were quite common. Today the dominant size is 275/80R22.5 and smaller (Figure 11.5). The incentive to downsize truck tires is to lower the platform of the trailer, thereby increasing the trailer volume, so that more freight can be carried, assuming permissible gross vehicle weights are not exceeded (7, 8, 9). In turn, the limiting factor here would be the docking height setting for loading and unloading. Also, lowering the height can allow the vehicle to pass under older low-level bridges not possible otherwise.

On-going technology needs will continue to include:

1. Decreasing rolling resistance. As mentioned earlier, as little as a 3% reduction in whole-tire rolling resistance can generate a 1% reduction in fuel consumption

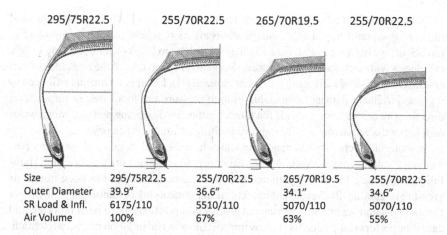

Size	295/75R22.5	255/70R22.5	265/70R19.5	255/70R22.5
Outer Diameter	39.9"	36.6"	34.1"	34.6"
SR Load & Infl.	6175/110	5510/110	5070/110	5070/110
Air Volume	100%	67%	63%	55%

FIGURE 11.5 Downsizing of Truck Tires (4, 5)

2. Global wear resistance, thereby increasing the miles or kilometers to removal, will always be a continuing need

3. Irregular wear resistance: this causes early tire removal and the new compliant suspension systems, developed to improve driver comfort, can tend to exacerbate this problem

4. Retreading: tire casings are an asset and research will continue to improve the durability of the tire. Environmental drivers also influence this such as reductions in scrap tire piles, materials conservation, and cost savings

5. Electronic device installation on tires to collect information on revolutions, mileage, operating temperatures, and inflation pressures. Though implementation of tire pressure monitoring systems for passenger car tires has been successful, driven by regulations, more progress is still needed for the commercial tire market.

Electronic devices built into tires, frequently abbreviated as RFID or radio frequency identification technology, would consist of several components:

- Power source, either built-in (preferred) or powered by a temporary magnetic field
- Antenna transmitter
- Processor
- Transducer (pressure measurement)
- Temperature measurement
- Revolution counter (for mileage)

Though large devices have been successfully installed in the large off-road tires, more work is required to develop comparable capabilities for highway truck tires (10).

In North America and Asia, the growth in the use of wide-base super singles for heavy-duty trucks, once considered promising due to improved fuel economy in the order of 4%, has abated for several reasons. New dual tire configurations have improved to the degree that the rolling resistance difference from singles has been closed, and dual tire configurations on dual axle configurations are more reliable, which is important in the US, unlike in Europe, where trailers may have three axles. As with increasing weight of vehicles with corresponding increasing load-carrying capacity of the tires, the counter concern is the risk of increasing pavement damage (11). Local governments will need to regulate pavement damage attributable to high-pressure tire footprints; in turn, this is directing tire design toward a wide flat footprint, thereby dispersing pressure over a wider area rather than a narrow one for improved rolling resistance efficiency.

In materials technologies, nanocomposites have been claimed to allow many performance improvements, such as fuel economy, tread wear performance, and durability. In practice, the only significant benefit to be demonstrated has been in tire air pressure retention (12). Significant reductions in permeability of nanocomposite rubber compounds, and specifically in elastomer–clay compositions, have been reported and can offer performance benefits such as improvement in tire inflation pressure retention.

11.4 FARM TIRES AND TRACKS

The concept which could become the greatest disruptor in the agricultural sector is tracks replacing tires (Figure 11.6). Tracks offer a number of advantages in off-road farming and light construction applications,

1. No punctures
2. Improved flotation
3. Less soil compaction, very important in areas with loose soils
4. Higher traction efficiency on softer soils (longer footprint)
5. Ride quality in rough fields, important in meeting productivity needs and when equipment is operated in 12-hour shifts
6. Side-hill stability and safety
7. Versus steel tracks, rubber tracks have no linkages & hinges thereby easing maintenance
8. Highway service (no pavement damage)
9. Less noise

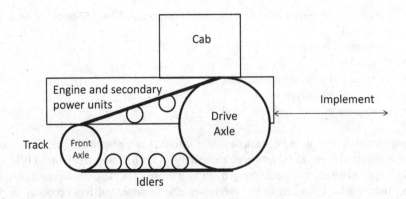

FIGURE 11.6 Tracks Replacing Tires

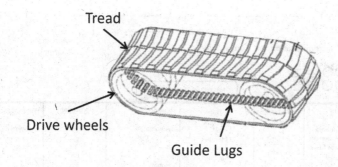

FIGURE 11.7   Track Construction (13, 14)

Tracks are constructed using steel cables, similar to what are used in steel cord-reinforced conveyor belts, which would have a steel cord wire coat compound similar to what is used in tires. The tread lugs are made of a highly fatigue-resistant compound and the guide lugs which help keep the moving track on the wheels have a very high hardness (Shore A Hardness) and a high tear strength natural rubber-based compound with reinforcing resins (Figure 11.7).

In addition to tracks, another potential disruptor could be steel ply radial tires for farming applications, made possible by adapting off-road tire technology. The flatter tire footprint can allow improvement in wear performance, assuming that any issues on soil compaction are mitigated.

11.5   OFF-ROAD TIRES (OTR)

Tires used off-road, such as in excavation and quarries, also continue to undergo changes. For definition purposes, OTR tires are considered to range from size 14.00R24 and upwards to for example 40.00R57 and 55/80R63. The largest claimed tire built for operational service is a 59/80R63 which is reported to have a section width of 1.47 m and a nominal load rating of 101 metric tons (7). The load-carrying capability of such tires would be expected to increase further as the load-carrying capability of the trucks also continues to increase. The limitation, however, will be transportation. The tires can be up to 4 m in height, thereby limiting their ability to be transported horizontally and, due to bridge heights, preventing vertical transportation. Though 67" tires have been developed, their production could be hindered without technology such as on-site two-piece tire assembly, pioneered by The Goodyear Tire & Rubber Company.

11.6   MANUFACTURING

Several trends in tire manufacturing are occurring. At the operational level, the engineer in the factory will have an on-going need for improvement in four areas:

1. Quality, both in raw materials, mixed compounds, and the final product
2. Cost, not so much as cost reduction which inevitably is linked to quality reduction, but cost containment and control of cost-creep

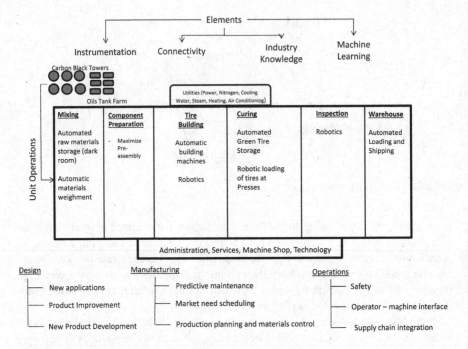

FIGURE 11.8 Tire Industry 4.0 Framework

3. Productivity and factory throughput rate
4. Minimization of the environmental impact, i.e., elimination of post-industrial waste and minimization of VOC emissions

Regarding productivity and increases in factory throughput rates, the primary opportunity is in equipment capacity and automation (Figure 11.8). Examples of such trends would be:

1. Increasing size of internal mixers, allowing more mixed compound
2. Pre-assembly of components made possible by multiplex extruders and calendar lines in parallel
3. Automated building machines.
4. Robotic handling of tires from curing through to shipping
5. Automated inventory control of compound, tire components in the Component Preparation department, and scheduling in Building and Curing operations

For passenger tires, there are three fundamental types of building machines. These are two-stage machines with two operators and where, at the 2nd stage, the belts and tread are applied and nominally produce 300 tires per shift; single-stage machines, with one operator, which nominally produces 400 tires per shift; and automated machines, with one operator to supervise between six and eight machines, and nominally produce around 500 tires per shift. There are several factors controlling an automated machine's output, such as machine parts operational reliability, component supply capability, loading of component reels onto servers, and tire specification run length (the lower the number of tires to be built and the shorter the run length, the greater the number of change-overs).

Automated plants are very good at producing a large number of tires of one specification. However, there is a challenge with respect to managing shorter build lengths, more changes in equipment settings, and the consequent decrease in total plant output. The overall strategy being developed to meet this challenge comes under the umbrella of Industry 4.0, loosely defined as the trend towards automation and data exchange in manufacturing technologies and processes, which include cyber-physical systems (CPS), the internet of things (IoT), industrial internet of things (IIOT), cloud computing, cognitive computing, and artificial intelligence. The elements of these systems are:

- Real-time supply chain management
- Supplier integration
- Accelerated prototyping
- Output versus demand forecasting
- Process improvement and predictive maintenance
- Quality and defect tracking

At the operational level, this infers adoption of tools such as the "internet of things", advanced robotics, and artificial intelligence, facilitating movement of components through the production process. Thus, as the number of tire specifications in production increases, the number of compounds in inventory will increase and, depending on the accounting procedures, plant operating costs will also increase. Industry 4.0 can provide a framework to manage increased complexity and costs, and still meet the market's needs.

Thus, in practice, up to the year 2025, Industry 4.0 infers:

- Dark factories, already seen in raw materials inventory areas and final tire storage prior to shipping
- More automated raw materials weighing and mixing, such as where each line will still have potentially two operators controlling the mixer but only one operator managing the roller-head extruders and lay-down of up to eight lines (compared to conventional lines with five operators per mixer)
- Component pre-assembly and automated component movement
- Automatic, robot-controlled building machines
- Robotic control and loading of curing presses
- Robotic sorting and loading in Shipping.

To date, mixing extruders and extrusion of components directly onto the building drum has largely been unsuccessful due to the inability to achieve the necessary compound uniformity and homogeneity. However, it remains an area of considerable research. Thus, any additional contribution of Industry 4.0 will be in the area of data management and process control.

In the future, 3-D printers, which build a three-dimensional object from a computer-aided design model, usually by successively adding material layer by layer ("additive manufacturing") is another potential disruptor. However, though printer-produced resolution is sufficient for many applications, the greater accuracy necessary for tire components may only be achieved by printing oversized versions of the component at conventional resolution, and then removing material using a higher-resolution subtractive process. With the current state of technology this could be cost prohibitive.

Operationally, the manufacturing organization can expect to see further changes in production planning and materials control. Efficiencies in this area could be an important differentiator, and systems such as Enterprise management developed by software companies such as SAP and Oracle, help with this. Industry 4.0 systems would be expected to further optimize manufacturing operations with the benefits of being in inventory management, scrap and waste reduction, and a plant operational environment.

It is anticipated that the net size of the labor force would remain similar to that of today. However, the demographics would shift, with greater numbers of skilled technicians participating in plant operations. Governments look on this favorably as a source of skilled employment and will thus offer industrial incentives to promote the industry, which would be in contrast to other industry segments which are becoming much less labor intensive.

11.7   COMPETITIVE PRODUCT BENCHMARKING

The monitoring of competitors' products is an important activity in ensuring a company will not get blind-sided by an unanticipated leap in a competitor's product performance or their launch of other new products. Benchmarking would occur at two levels, firstly, at the tactical level, where competitors' new products are routinely monitored, and secondly, at the strategic level, which would require collection of more in-depth technical intelligence. In both cases, the highest ethical standards must be followed. In adhering to such standards, competitors' products can be purchased on the open market and knowledge of activities can be obtained from trade journals, conferences, publications, and patents.

At a tactical level, benchmarking of competitive products, upon their purchase in the open market, consists of three elements, namely design, construction, and materials analysis. Two institutes which have a pro-active program of benchmarking are Smithers Scientific Services, which is based in Akron, Ohio, and the Beijing Rubber Development Institute. Tires which have been acquired on the open market are studied as follows;

1. Design; tread patterns, net footprint contact area, whole tire weight, and in some instances, tire testing such as durability, rolling resistance, and recently, tire inflation pressure loss rates
2. Construction: all component gauges, wire and fabric construction, ends per inch (EPI), component dimensions, and lay-ups
3. Materials analysis: frequently based on thermogravimetric analysis but additional parameters, such as carbon black, have also been estimated with a high degree of accuracy

4. Laboratory tire tests, such as rolling resistance.

It is relatively easy to measure the component dimensions of a cut tire and to copy tread patterns, thereby duplicating a competitors' product for internal benchmarking. However, compound replication is considerably more difficult. There are three levels of analysis: i) preliminary thermogravimetric analysis (TGA) to identify the ratios of the ingredients in a compound (Figure 11.9), ii) polymer identification, volatile content, or identification of low-molecular-weight products and by-products from vulcanization reactions, using Fourier-transform infrared spectroscopy (FTIR)and chromatography techniques, and iii) analysis of carbon black by microscopy and ash identification. Thermogravimetric analysis or TGA may also be used to make an initial identification of polymers in a compound containing blends (Table 11.6).

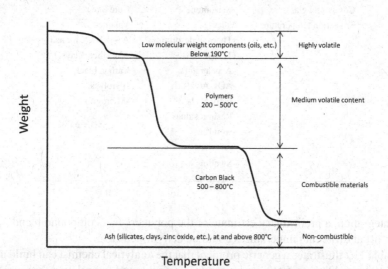

FIGURE 11.9 Format of a TGA Analysis (17) ASTM E1131 Standard Method for Compositional Analysis by Thermogravimetric Methods

TABLE 11.6
Polymer Identification Using TGA (15)

Polymer	Temperature (°C)
EPDM	461
NR	373
Polyisoprene	373
Solution SBR	445
Emulsion SBR	442
Polybutadiene	461
Butyl	386

TABLE 11.7

In-depth Analysis of a Rubber Compound (2)

	Depth	
1	**2**	**3**
TGA	TGA	TGA
1 Polymer type	FTIR	FTIR
2 Carbon black level	Polymer types	Pyrolysis
3 Ash (silica, clay, ZnO)	Polymer ratios	^{13}C NMR
4 Total volatiles	Pyrolysis	-Microstructure
	-% styrene	Ozonolysis
DSC	-% butadiene	-Full sequencing
Compound glass	-% isoprene	Cure State
Transition temperature	Volatiles	Volatiles
	-Thermal desorption	AA. EDAX for ash
	(TD-GCMS)	-Zn, silica, clays, talc
	-Accelerators	Carbon black
	-AOs, AOz's	-Pyrolysis
	-Fatty acids	-Micrographs
	-Resin residues	
	-% sulfur	
	Cure state	
	-Swelling	

By integration, a preliminary estimate of the polymers in a compound blend which come off at different temperatures could be determined (15).

Table 11.7 illustrates a generic protocol that the analytical chemist can build upon, ultimately enabling an attempt to reproduce a compound formulation. Frequently, a TGA is sufficient to identify a compound composition, e.g., a TBR all-natural rubber tread compound with nominally 50 phr carbon black, ~5 phr of process oil, and semi-EV vulcanization system.

A program can then be started where a tire, such as the premium steer axle truck tire, is acquired periodically, say every 6 months, analyzed and then, after a period of time, should an anomaly in an analysis appear, a more in-depth analysis would follow (Table 11.7). The analysis would also include a measurement of the compound mechanical properties, though, due to the skill of the operators and their ability to extract adequate samples for testing, this might be limited to components such as treads, inner liners, and sidewalls (Table 11.8).

One gap in both the rubber compound development and benchmarking has been the adoption of combinatorial chemistry and other high throughput methodologies (16). Due to the complexity of compound formulations and the diversity of test methods and equipment used to characterize a compound, automated test protocols have yet to emerge. However, with the ongoing evolution of equipment, such as the Rubber Process Analyzer (RPA)- coupled with expansions of the use of multivariate data

TABLE 11.8

Test Protocol for Tire Components (2) (Assuming that they can be extracted in adequate quantity)

Level		
1	2	3
Tensile strength	Tensile strength	Tensile strength
Rebound	Rebound	Rebound
Tear strength	Tear strength	Tear strength
Abrasion	Abrasion	Abrasion
Specific gravity	Specific gravity	Specific gravity
Dynamic properties [G', G'', tan delta]	Dynamic properties [G', G'', tan delta]	Dynamic properties [G', G'', tan delta]
	Temperature sweeps	Temperature sweeps
	Adhesion	Adhesion
	Modulus profiling	Modulus profiling
		Reverse engineer
		Mechanical properties match

analysis and neural networks, it could be envisaged that such methodologies could be found in future rubber technology and engineering laboratories.

11.8 COMPETITIVE TECHNICAL INTELLIGENCE

Product benchmarking should be an integrated part of a competitive technical intelligence program. In addition, successful competitive technical intelligence programs are, in many instances, a collaborative effort between Technology and Marketing (18). The work can then take two tracks, one is routine benchmarking, whereas the second focuses on specific targets. At a strategic level, review of future trends in the tire industry is incomplete without Marketing's discussion of competitive threats. Innovations in designs, which take one of two forms, namely those novel tire launches designed to generate marketing attention and indirectly, sales, and prototypes being developed to go into production. For example, novel designs are regularly featured in European car magazines which, in addition to allowing drivers to experience tire–vehicle performance enhancements, also allow favorable tire company publicity. Regardless, monitoring of new innovations is essential for a variety of reasons:

1. Avoidance of being blind-sided
2. Increasing external focus and awareness
3. Acquisition of new technology
4. Input for a component of a tactical project plan, i.e., the design input step required under QS9000 and TS16949 development protocols (Figure 10.14), and which is also an integrated part of other development tools, such as the "House of Quality" and "Quality Function Deployment" (Figure 11.10)

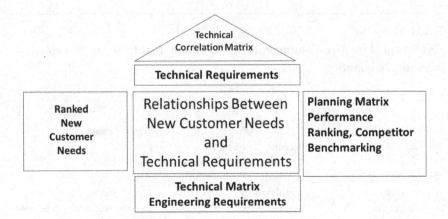

FIGURE 11.10   Quality Function Deployment and House of Quality (19)

5. Component for a strategic business technology plan, i.e., competitive bench-marking is a critical element in any business plan
6. Anticipating or countering competitive threats
7. Cost savings in research and development
8. Product positioning
9. Motivating and focusing technology teams and management

Those organizations with effective competitive technical programs could expect to see new product launches proceeding more smoothly. Sources of market and technical technology are obvious in that they can be readily listed;

1. Literature
2. Patents, *via* profiling a competitor's activity, volume, and trend analysis, and others who might also be engaged in such activity
3. Conference papers
4. Sales and marketing input
5. Regulatory authorities, such as the Environmental Protection Agency (EPA)
6. Media, such as newspapers and specialist magazines
7. Competitors' products, both from customer feedback and reverse engineering
8. Strategic suppliers, through sharing of their internal research and development, and corresponding market interest.

The gate keepers for such activities are the senior researchers in the technical organization because this group can recognize what is significant and what is not.

Suppliers similarly can play a critical role in benchmarking. Strategic suppliers offer the following advantages:

• Access to capable research and development teams
• Laboratory resources, enabling reductions in project cycle time

- Information technologies
- Competitive information, global footprint, and incentive to succeed.

Furthermore, they are committed to their strategic customer's success, thus opening up opportunities to the second-tier manufacturer for favorable pricing, risk-sharing or mitigation, and will have a longer-term business horizon. Thus, companies successfully executing competitive technical intelligence programs will also demonstrate integrated supplier project collaboration and management activity, which enhances the tire company's competitive position. The critical enabler for such a relationship, however, is the Confidentiality Agreement, and there are two elements to this, i) the legal document, and ii) the reputation of the supplier to honor the spirit of the agreement, and *vice versa*.

11.9   CONSERVATION AND RECYCLING

There are several reasons to recycle or reclaim rubber materials from tires. Environmental protection is the greatest driver and, given the excess piles of waste tires accumulated in tire dumps or post-consumer waste, is a factor of considerable importance. The reduction in the use of carbon-based materials is a second area which is attracting attention. A third area is that the automotive industry has attempted to set targets for recycled content of 25% of post-consumer and post-industrial scrap into their products with no increase in cost or loss in performance. Post-consumer scrap recycling covers the re-use of products which have completed their service life by either returning tires to service through retreading or by using them in other applications. Other applications of products include grinding them into a powder or returning them to their original state *via* a devulcanization process. Post-industrial scrap is the use of waste material generated in the original manufacturing process. In this instance, the goal is to ensure that all material used in the manufacture of a product is converted to products of high quality (3). Recycling, better described as conservation, has been described as falling into one of four categories, namely materials reduction, reuse, true recycling, and reclaim.

11.9.1   REDUCTION

Materials Reduction efforts have focused on optimum use of materials, weight reduction of tires or other engineered product, and product component gauge optimization. This has largely been facilitated through new manufacturing systems and new designs. Techniques such as electron beam radiation of tire components help in this area.

11.9.2   REUSE

Reuse of tires and other industrial rubber products has been directed either for use as fuel or for retreading. Excluding those tires going to landfills, stockpiles, or other storage facilities, most are used for fuel. Retreading of aircraft and commercial truck tires is probably the most ideal use of worn products. In the case of commercial aircraft tires, up to four and sometimes five retreads are possible. For commercial truck tires from size 9.00R20 up to 12.00R24 or 315/80R22.5, two retreads are not unusual.

11.9.3 RECYCLE

The major methods for recycling of existing rubber are ambient grinding, cryogenic grinding, and wet grinding. Depending on the final particle size, resulting products are useful for controlling compound cost and improved processing when added to new compounded rubbers.

11.9.4 RECLAIM

Reclaim of rubber refers to the recovery of original elastomers in a form that can be used to replace fresh polymer. There is, again, a range of techniques available to produce such materials:

a. Ultrasonic Devulcanization: though it has not been achieved commercially, it continues to be a potential method to allow reclaim of the original polymer. Sulfur–Sulfur bonds have lower bond energy than carbon–carbon bonds. Given this, ultrasonic waves can have enough energy to selectively break the sulfur bonds, thereby devulcanizing the compound.

b. Chemical Devulcanization: such methods involve mixing rubber peelings in a high swelling solvent and a catalyst. After heating, there is significant reduction in cross-link density. Though other chemical techniques are being investigated, any future system will most likely involve a catalytic degradation in a solvent at high temperature and pressure.

c. Thermal Devulcanization: involves use of microwaves, inducing an increase in temperature with preferential breaking of sulfur–sulfur bonds. Due to the cost of operating such systems, there has been no successful commercial system, but pilot plant facilities have been in operation.

d. Chemo-mechanical and Thermo-mechanical Devulcanization: Such systems have ranged from the simple addition of vulcanization system ingredients to crumb rubber, or polymer surface modification, thereby adding functionality to the surface of the particles, to treatment at higher temperatures with the same intent of activating the surface. No commercially successful systems have been developed, though pilot plant facilities have been developed.

Reclaimed rubber was used in significant volume through the early 1970's in the US. However, the growth of the radial tire, environmental regulations, and the expansion of SBR and BR manufacturing, resulting in low original rubber prices, caused significant contraction of the reclaim rubber industry. In the 1960's, there were estimates of as much as 270,000 tons per year of reclaimed rubber being used in the US. In the year 2002, it was estimated that 27,000 tons may have been used. This consisted mostly of reclaimed butyl compounds used in some tire inner liners and the

TABLE 11.9
Tentative Guidelines for Use of Recycled Materials in Tires (2, 9)

Component	Passenger Tires (PCR)	Light Truck Tires (RLT)	Commercial Truck Tires (TBR)	Retreading (TBR) and Retreadable Casings
Treads	Yes	Yes	No	Yes
Sub-tread	No	No	No	Yes
Casing, ply	No	No	No	No
Bead filler	Yes	Yes	No	No
Sidewall	Yes	Yes	No	No
Wedges	Yes	No	No	No
Barrier	Yes	Yes	Yes	Yes
Inner liner	No	No	No	No

barrier with reclaimed NR being used in mats and low-end static applications, and reclaimed silicone compounds being used for automotive and electrical applications.

In processing post-industrial scrap back into original compounds, the manufacturing engineer would have a defined specification showing what can be recycled, and how much can be added, without causing any deterioration in the mechanical properties of the tire compound. Table 11.9 provides a starting point or guideline to develop recycle specifications. For example, it would be suggested that the inner liner would not have any recycled content, in order to minimize the risk of fatigue-related cracks or deterioration of other important properties. In all cases the components containing recycled compound and the amount of recycled content would be determined experimentally, and after field trials to validate the data generated data, so as to ensure that there were no adverse effects on performance.

Finally, nitrogen inflation has attracted attention as a means to enhance tire performance and service life. The air in our atmosphere is nominally 80% nitrogen so claims of improved performance of tires with nitrogen inflation rest on the assumption that there is no oxygen and trace carbon dioxide in the inflated tire. Pure nitrogen inflation in place of air is claimed to:

- reduce inflation pressure loss rate
- potentially improve fuel economy, due to stable inflation pressures
- improve laboratory tire durability improvement (1.7-m test wheel)

Though the claims are genuine, it requires pure 100% nitrogen and, by inference no O_2, water vapor, or other potential contaminants. Nitrogen inflation is of considerable importance for large off-road tires, such as 40.00R57 or the new, larger 63" diameter tires, as well as aircraft tires. Nitrogen inflation is important for aircraft safety and reduction with respect to the risk of auto-ignition or other fires due to any over-heating conditions that may arise. Similarly, large earthmover tires are also susceptible to auto-ignition.

For automobile tires, nitrogen would be produced from compressors, and, in this instance, depending on the condition of the equipment, up to 5% of the inflation gas may still consist of oxygen. Oxygen will permeate through a tire at a faster rate than nitrogen. It has also been reported that, over time, as the tire is rechecked and inflation pressure topped up, there will be a shift in the ratio of oxygen to nitrogen such that, after 3 to 4 years, the nitrogen content will tend toward 90% to 95% of the air in the tire. In consequence, end-users may not see any improvement in tire performance. In practice, to achieve the claimed improvements due nitrogen inflation, a pure source of nitrogen would be needed, i.e., from liquid nitrogen, though costs and infrastructure could render this strategy prohibitive.

Regardless of air or nitrogen being used as the inflation medium, pressure retention is still of paramount importance and will thus necessitate a full 100 phr halobutyl and preferably bromobutyl inner liner compound.

11.10 SUSTAINABILITY

The area of sustainability, which is much more diverse than environmental protection practices, would include raw materials technologies, manufacturing efficiencies, materials conversion, and product performance (20). It is therefore a multi-dimensional issue, covering up to ten elements:

1. Safety
2. Water conservation
3. Power and energy efficiency
4. Emissions
5. Materials conversion efficiency, with 100% being the ideal
6. Equipment reliability, preventative maintenance, efficiency, and up-time
7. Biomaterials
8. Recycling or disposal of post-industrial scrap such as compounds, reinforcements, and tires not passing final inspection
9. Secondary pollution controls such as disposal of pallets and cardboard packaging, excess chemicals inventories, toxic materials handling, mold flash and trim from production
10. Social awareness, employee involvement, community commitment, and the enterprise's contribution to society.

Future trends should focus more on these areas as opportunities, if for no other reason than that the resulting cost savings can be substantial. There is considerable publicity around the topic of sustainability but, to have any relevance at the manufacturing or operational level, the focus would need to be on water and power efficiencies, emissions, waste elimination, and employee safety.

For example, water is used in cooling operations, heating, and steam for curing processes. As much as 12,000 liters of make-up water can be used per ton of finished product, representing a significant cost. Collection and treatment of wastewater can enable this to be lowered to near 7000 liters. Power consumption between tire factories of similar capacity can also vary widely.

A further aspect of sustainability is product-life extension, with tire retreading falling within this scope. The retread industry in North America and Europe has declined since 2016, due to increasing Asian manufacturing capacity and unfavorable cost structure. This is not expected to continue. When the upturn does occur, end-users will expect two or more retreads per casing and, therefore, future development work on casing durability is necessary. Pre-cure appears to be the most popular method of retreading and it is expected to continue.

Safety is a common characteristic in well-managed factories producing high-quality tires. Systems put in place to ensure safe operations, by default, ensure high quality due to factors such as better equipment reliability, better process monitoring, and efficient functioning of utilities (e.g., no leaking steam lines). Other, intangible benefits can also be available, such as employee engagement which has a further indirect impact quality. One of the best procedures for ensuring company safety, which is part of a manufacturer's sustainability program, is the employee "behavioral-based safety observation" used extensively in the oil and gas industry. Such protocols contribute not only to a safe working environment but also provide a means to quickly identify and report operations drifting out of specification and identify opportunities to further improve operations.

11.11 SUMMARY

The most important point in summarizing future trends in tire engineering and technology is that it is expected that the current pneumatic tire will still be in production ten years from now. However, there may be some modifications, as the technology is adapted for new vehicle platforms and ongoing regulatory demands. Although many original tire technology concepts have been developed over the past thirty years, few, if any, will make it into production. However, concept development is an essential activity for many reasons such as for marketing, prototyping, and as a problem-solving tool. Though no new raw materials will be expected, new low-cost feedstocks and bio-sourced materials from which new molecules could be developed are potential disruptors.

In manufacturing, it would be expected that there will be consolidation in the China tire industry, something that is being driven by the central government, and, in parallel, a drive to improve quality and meet global performance expectations, such as uniformity and rolling resistance. Industry 4.0 and all that implies will be another substantial revolution.

REFERENCES

1. Colvin H. General Purpose Elastomers. In *Rubber Compounding, Chemistry and Applications*, Ed. B Rodgers. CRC Press, Boca Raton, FL. 2015, 33–82.
2. Rodgers MB ELL Technologies LLC, Cedar Park, TX. 2020.
3. Rodgers B. Natural Rubber and Other Naturally Occurring Compounding Materials. In *Rubber Compounding Chemistry and Applications*. Ed. B Rodgers. CRC Press, Boca Raton, FL. 2015, 1–32.
4. D'Cruz B, Rodgers B, Sharma B. *Evaluation of Anti-degradant Systems for Enhancing Performance of Bromobutyl Based Tire Innerliners.* Presented at a Meeting of the American Chemical Society Rubber Division. Paper 18. Cleveland, OH. 2011.
5. SAE J1016. *Automated Driving. Levels of Driving Automation.* Society of Automotive Engineers, Warrendale, PA. 2014.

6. Self-Driving Cars. 2019. Wikipedia. https://en.wikipedia.org/wiki/Self-driving_car.
7. Gabor J, Wall J, Rodgers B. *Overview of Medium Radial Truck and Off the Road Tire Technology.* Presented at a Meeting of the American Chemical Society Rubber Division, Providence, RI. 2001.
8. Tire & Rim Assoc. *Year Book.* Copley, OH. 2006.
9. Rodgers B. *Commercial Truck Tire Technology Trends.* Presented at the Ohio Rubber Group, Akron, OH. 2010.
10. Belski G, Carmickle SP, Rodgers B. *Electronic Identification Technology for Radial Medium Truck Tires.* Presented at the International Tire Exhibition and Conference, Akron, OH. 1994.
11. Yap P. *A Comparative Study of the Effect of Truck Tire Type on Pavement Contact Pressures.* SAE Technical Paper Series 881846. Warrendale, PA. 1988.
12. Rodgers B, Webb R, Wang W. *Advanced Tire Innerliners.* Presented at a Meeting of the American Chemical Society Rubber Division, Pittsburg, PA. 2005.
13. Rodgers MB, Krishnan RM, Sandstorm PH, Maly NA, Gordon LA. *Endless Rubber Track and Vehicle Containing Such Track.* US Patent 6296329. 2001.
14. Krishnan RM, Lukich LT, Rodgers MB, Beery RE, Rabatin GC. *Cold Environment Endless Rubber Track and Vehicle Containing Such Track.* US Patent 6799815. 2004.
15. Brazier DW, Nickle GH. Thermo-Analytical Methods in Vulcanizate Analysis II. Derivative Thermogravimetric Analysis. *Rubber Chemistry and Technology.* Vol. 48, pp. 661–677. 1975.
16. Czarnik AW, DeWitt SH. *A Practical Guide to Combinatorial Chemistry.* American Chemical Society, New York. 1997.
17. ASTM E1131-08. *Standard Method for Compositional Analysis by Thermogravimetric Methods.* American Society for the Testing of Materials. West Conshohocken, PA. 2014.
18. Coburn MM. *Competitive Technical Intelligence.* American Chemical Society and Oxford University Press, Washington, DC.1999.
19. Akao Y. *Quality Function Deployment.* Productivity Press, Cambridge, MA. 1990.
20. Bjacek P. Sustainability. In *Hydrocarbon Processing.* Ed. Lee Nichols. Catherine Watkins Publishing. 2020, Volume 99, No. 1, 33–46.

Appendix 1: Abbreviations for Tire Polymers and Polymer Additives

The International Institute of Synthetic Rubber Producers has prepared a list of abbreviations for all elastomers (3). For example, BR denotes polybutadiene, IR is synthetic polyisoprene, and NBR is acrylonitrile-butadiene rubber (Table A.1)

TABLE A.1

International Institute of Synthetic Rubber Producers Abbreviations for Selected Elastomers (3)

AU	Polyester urethane
BIIR	Brominated isobutylene-isoprene rubber (bromobutyl rubber)
BHT	Butylated hydroxytoluene. Antioxidant and protects against degradation.
BR	Polybutadiene
CIIR	Chlorinated isobutylene-isoprene rubber (chlorobutyl rubber)
CPE	Chlorinated polyethylene
CR	Chloroprene rubber
CSM	Chlorosulfonyl polyethylene
EAM	Ethylene-vinyl acetate copolymer
EPDM	Terpolymer of ethylene, propylene and a diene with a residual unsaturated portion in the chain
EPM	Ethylene propylene copolymer
ESBO	Epoxidized soybean oil. Epoxy groups capture any acids, including excess stearic acid. Concentration used is approximately 1.3%
EU	Polyether urethane
GPR	General-purpose rubber
HNBR	Hydrogenated acrylonitrile butadiene rubber
IIR	Isobutylene isoprene rubber
IR	Synthetic polyisoprene
NBR	Acrylonitrile butadiene rubber
SBR	Styrene butadiene rubber
E-SBR	Emulsion styrene butadiene rubber
S-SBR	Solution styrene butadiene rubber
X-NBR	Carboxylated acrylonitrile butadiene rubber
X-SBR	Carboxylated styrene butadiene rubber
Y-SBR	Block copolymer of styrene and butadiene

There are also a number of definitions which merit discussion.

Appendix 2: Recognized Industry Abbreviations for Accelerators

Abbreviation and Commercial Descriptions	Chemical Name	Function
Amylphenol disulfide 1	Amyl disulfide polymer (23% S)	Sulfur donor
Amylphenol disulfide 2	Amyl disulfide polymer (28% S)	Sulfur donor
Amylphenol disulfide 3	Amyl disulfide polymer (18% S)	Sulfur donor
Amylphenol disulfide 4	Amyl disulfide polymer (30% S)	Sulfur donor
Amylphenol disulfide 5	Amyl disulfide polymer (27% S,10% HSt)	Sulfur donor
BCI-MX	1,3-Bis(citraconimidomethyl) benzene	Reversion res.
CBS	N-Cyclohexyl-2-benzothiazolesulfenamide	Primary acc.
CTP	N-(Cyclohexylthio) phthalimide	Retarder
DBQDO	p-Quinone dioxime dibenzoate	Quinone cure
DBTU	Dibutylthiourea	Accelerator
DCBS	Dicyclohexylbenzothiazole sulfonamide	Primary acc.
DETU	Diethylthiourea	Accelerator
DOTG	Di-o-tolylguanidine	Secondary acc.
DPG	Diphenyl guanidine	Secondary acc.
DPPD	Diphenyl-p-phenylenediamine	Accelerator
DPTU	N,N'-Diphenylthiourea	Accelerator
DTDM	4,4-Dithiodimorpholine	Vulcanizing agent
ETU	Ethylthiourea	Accelerator
HTS	Hexamethylene-1,6-bis(thiosulfate) disodium salt, dihydrate	Reversion res.
MBS	Oxydiethylene benzothiazole-2-sulfenamide	Primary acc.
MBT	Mercaptobenzothiazole	Accelerator
MBTS	Mercaptobenzothiazole disulfide	Primary acc.
QDO	p-Quinone dioxime	Quinone cure
TBBS	tert-Butyl-2-benzothiazole sulfonamide	Primary acc.
TBSI	N-t-Butyl-2-benzothiazole sulfenimide	Primary acc.
TBzTD	Tetrabenzylthiuram disulfide	Secondary acc.
TMTD	Tetramethylthiuram disulfide	Secondary acc.
TMTM	Tetramethylthiuram monosulfide	Secondary acc.
ZBEC	Zinc dibenzyldithiocarbamate	Secondary acc.
ZBPD	Zinc dibutylphosphorodithiate	Accelerator

Abbreviation and Commercial Descriptions	Chemical Name	Function
ZDBC	Zinc dibutyldithiocarbamate	Secondary acc.
ZDEC	Zinc diethyldithiocarbamate	Secondary acc.
ZDMC	Zinc dimethyldithiocarbamate	Secondary acc.
ZIX	Zinc isopropyl xanthate	Low-temperature acc.

Appendix 3: Common Terms for Compounding Ingredients

Altax	Benzothiazyldisulfide
Barytes	Barium sulfate
DEG	Diethyleneglycol
DIAK No. I	Hexamethylenediamine carbamate
Di-Cup 40C	Dicumyl peroxide (409 active)
Escorez™ 1102	Petroleum-based resin
Flexon™ 876	Paraffinic mineral oil
Flexon™ 641	Naphthenic mineral oil
Flexon™ 580	Naphthenic mineral oil
Flexon™ 391	Aromatic mineral oil
MBI	Mercaptobenzimidazole
Mineral Rubber	Blends of maltenes, asphaltenes
Parapol 2225	Low-molecular-weight isobutylene–butane copolymer
Retarder W	Salicylic acid
Ultramarine blue	Blue pigment
Wood Rosin	Derivatives of abietic acid

Index

Printed in the United States
by Baker & Taylor Publisher Services